# Blends of Natural Rubber

# Blends of Natural Rubber

## Novel Techniques for Blending with Speciality Polymers

Edited by

Andrew J Tinker and Kevin P Jones

Suported by the Common Fund for Commodities

CHAPMAN & HALL

London · Weinheim · New York · Tokyo · Melbourne · Madras

**Published by Chapman & Hall, an imprint of Thomson Science, 2–6 Boundary Row, London SE1 8HN, UK**

Thomson Science, 2–6 Boundary Row, London SE1 8HN, UK

Thomson, 115, Fifth Avenue, New York, NY 10003, USA

Thomson Science, Suite 750, 400 Market Street, Philadelphia, PA 19106 USA

Thomson Science, Pappelalle 3, 69469 Weinheim, Germany

First edition 1998



Thomson Science is a division of International Thomson Publishing

I(T)P

Typeset in 10/12 Palatino by Scientific Publishing Services (P) Ltd., Madras, India
Printed in Great Britain by St Edmundsbury Press, Bury St Edmunds, Suffolk

ISBN 0 412 819406



A catalogue record for this book is available from the British Library

Library of Congress Catalog Card Number: 97-75075

# Contents

# Foreword

To those knowledgeable in the art, polymer blends is not a new subject. For the last twenty-five years it has been one of the most treated topics of pure and applied polymer science. The reasons for this continued interest are mainly the following points:

- improvement of impact resistance of thermoplastics and thermosets
- combination of properties of both (or more) partners
- structural elucidation of compounded materials with the progress of modern analytical techniques
- detection and role of phase boundaries and their influence on compatibility
- cost savings for expensive engineering materials.

In the realms of elastomers, on the other hand, blends are indeed much older than in the above quoted plastics area – ever since the advent of synthetic elastomers after the First World War we have tried to stretch supplies of natural rubber (NR), in particular since its property levels were initially vastly superior to the synthetic counterparts.

This has certainly changed during the last forty years and when MRPRA proposed to the CFC a three-year research programme on NR blends, it was geared to foster an increased utilization of NR by penetrating into new markets. Now one should not believe that the four blend families which are outlined in this book are entirely newly invented combinations. Rather, all of these copolymers, *viz*. NBR, EPM, EPDM and ENR, have been proposed for combination with NR somewhere in the scientific or patent literature.

The problem associated with such combinations is that usually the property levels of a single species are not maintained in a blend, but are below expectations; there are rare cases of synergism. It is here that the expertise of a world-renowned institute like MRPRA comes into play and it is their basic polymer science, the analytical skills, the compounding know-how and the processing experience which lead to interesting new products.

As one reads the book, one must arrive at the conclusion that the CFC was well advised to confer this project to IRRDB, INRO and their fellows. Although not every question has found a complete answer, the blends project might be called a great success.

Prof. Dr Hans Schnecko
Editor in Chief
*Kautschuk Gummi Kunststoffe*

# Preface

The function of this brief Preface is to state the primary aim of the work and to acknowledge the providers of funding, expertise and other assistance. The principal objective was, and is, to increase the market value of natural rubber, and thereby enhance the income of the producers of natural rubber, the majority of whom are small farmers with low incomes living in the classical rubber-growing areas of southern Asia and west Africa. There is nothing novel about this aim; for most of this century the producers of natural rubber have sought to increase their earnings by enlarging the markets for their commodity. A significant secondary aim was to transfer new techonology to these same rubber-producing countries to enable them to develop downstream manufacturing with appropriate materials, that is by limiting the amount of material which needs to be imported.

The Common Fund for Commodities (CFC), who provided about half of the funding, and the Malaysian Rubber Research and Development Board (MRRDB) were the two major agencies involved, although it must be stressed that other organizations had significant roles to play. The Common Fund for Commodities was established in 1989 to assist in the development of a wide range of materials. The Fund provides loans and grants to finance research, development and marketing projects. The Blends Project which forms the basis for this book was one of the CFC's first projects. It is generally agreed that the project was highly successful in all its aspects, especially in that it was able to be completed within the budget set for it.

It is a requirement for the CFC to seek projects which will bring benefits to a group of countries rather than to a single nation, and it is required to operate through an International Commodity Body. In the case of the rubber industry there are two: the International Rubber Study Group and the International Natural Rubber Organization (INRO). The latter agency was involved in the Blends Project. The primary task of INRO is to operate the International Natural Rubber Agreements which are designed to stabilize natural rubber (NR) prices to ensure acceptable producer incomes and to enable consumers to be able to predict future costs.

The International Rubber Research and Development Board (IRRDB) is a far less formal organization than the bodies considered so far. The IRRDB was established in the late 1930s to coordinate research and development work. It was reorganized in 1961 and currently has Member Institutes in fifteen NR-producing countries. In the case of the Blends Project it formed a part of the channel for communication between the CFC, INRO and MRRDB, and also acted as the major vehicle for technology transfer.

The bulk of the work was performed at the Malaysian Rubber Producers' Research Association (MRPRA), which is a unit of the MRRDB. Since the end of the project, MRPRA has changed its name to the Tun Abdul Razak Research Centre. MRPRA provided the scientific and technological expertise and a great deal of time; in broad terms the cost of the project was borne equally by the MRRDB and the CFC.

The IRRDB organized a programme of fellowships for Member Institutes in China, India, Indonesia, the Philippines, Sri Lanka and Thailand. The Fellows each spent five months at MRPRA and were integrated into the research programme: their contribution is self-evident from this book. They were not mere passive observers, but for a time became a part of the research team. It is believed that this aspect of the project may be worthy of study by other organizers of programmes which are seeking to optimize the transfer of technology.

The majority of those involved are contributors to this book, and it is hoped that authorship will be sufficient to demonstrate recognition of their work. On the other hand, there were several major facilitators who would remain unrecognized unless mention is made of them here. Dr Peter Allen, the former Secretary of the IRRDB, had a major coordinating role and ensured that the project remained firmly within its budget. He greatly assisted with assembling the Fellows from across the world.

Mr Christopher Goldthorpe of INRO and Mr S. Olowude of the Common Fund were closely associated with the project. Prof. Dr Hans Schnecko was a consultant to the project and has contributed an informative foreword. The project leader, Dr Andrew Tinker, is quite properly and clearly both the senior editor and an author, but it would be unjust not to recognize the role played by Dr Cris Baker, the Director of the Tun Abdul Razak Research Centre (TARRC) where the bulk of the work was performed and Datuk Aziz, Director of the Rubber Research Institute of Malaysia.

The editors would like to acknowledge the considerable assistance given by Dr Stuart Cook in the task of editing. The special contributions made by the Physical Testing and Materials Characterization Groups at

TARRC must be recorded, and in the case of the latter the expertise provided by Kevin Jackson, Barry Gilbey, Katherine Lawrence and Christine Lewan. Jayne Freund, who provided technical assistance to the research team, deserves special recognition. Recognition also needs to be made to the Power Synthetic Rubber Manufacturing Corporation in the Philippines who released Rose Escolar for a lengthy secondment to the project. The project concluded with a Workshop in Penang in 1995 which afforded the only opportunity for all the participants to meet (as the Fellows had been limited to pairs at any one time). The book has grown from the papers presented at this Workshop, and it is hoped that those who had hoped to read the proceedings earlier will consider that the more balanced approach which a book provides will more than compensate for any delay.

Kevin Jones
Secretary
*International Rubber Research and Development Board*

# 1

# Introduction – the book and rubber blends

*Andrew J. Tinker*
*Tun Abdul Razak Research Centre, MRPRA, UK*

The origins of this book are described in the Preface and are important in defining the form which it takes. It is neither an exhaustive treatment of the subject – rubber blends – nor one of the oft-encountered collection of seminar papers. Whilst the chapters have a number of authors, all have worked as a team tackling a number of related topics in a defined programme of work. Several consequences flow from this. There are common threads and a generality of approach. Only certain blend systems are considered, even where an approach might logically be also applicable to another combination of elastomers. The work described was also constrained by falling within a fixed time frame, and scope for further advances remains in many of the more practical areas.

Some essential background to both the subject – rubber blends – and the approaches adopted in endeavouring to make advances across quite a wide front are given here. The wider applicability of some of the solutions developed in this work is also highlighted.

The following two chapters (2 and 3) then describe characterization techniques which are central to the work on blends and so provide the background to experimental results presented in ensuing chapters. The procedures covered in these two chapters were established before the work on blends presented here was initiated. The next chapter (4) also considers a characterization technique, but describes a new approach undertaken as part of the programme of work – an attempt to extend the principles of measurement of crosslink densities in vulcanized blends by

*Blends of Natural Rubber.* Edited by Andrew J. Tinker and Kevin P. Jones.
Published in 1998 by Chapman & Hall, London. ISBN 0 412 819406

swollen-state NMR spectroscopy, described in Chapter 2, to solid-state NMR spectroscopy.

The following chapters (5 to 17) then deal with the various blend systems studied, generally taking the pattern of one or more chapters dealing with the more basic issues associated with the particular blend system followed by one or more chapters relating to applying advances made within the context of specific applications. The chapters are essentially synoptic, describing the advances made in sufficient detail to inform, but without presenting all of the work and experimental detail. That larger task is being undertaken through the normal process of publication of a large number of detailed papers in appropriate scientific journals.

Given the large number of elastomers, rubber chemicals and instrumental techniques mentioned in the book, a glossary is provided at the end as an aid to those new to the subject.

## 1.1 SOURCES OF PROBLEMS – AND SOLUTIONS

The use of blends of elastomers is almost as old as the synthetic rubber industry and generally stems from an understandable desire to combine the best features – technical or economic – of two elastomers. In some instances, this desirable outcome has been attained relatively easily to the benefit of the industry and the end user – blends of the general purpose rubbers in tyre treads, for example. In other instances more guile has been required for a successful outcome, and there are a good number of desirable combinations which have not yet proved successful. Blends of general purpose rubbers with special purpose or speciality elastomers often figure in this last category.

A good starting point is to consider first the factors which are peculiar to elastomer blends and which are important in governing the properties of the vulcanized blends. The scope for problems is then apparent – but the same scope is available for control. These points are discussed within the context of the blends considered in the following chapters. The relevance of these studies to real industrial problems is made clear. Finally, the wider implications of the developments described here are explored.

### 1.1.1 Factors affecting blend properties

In addition to the normal considerations of compounding a vulcanizate of a single elastomer for a particular end use – such as filler type and loading, type of cure system and level, and antidegradants – there are a number of factors which arise in blends of elastomers, primarily because these almost always have more than one polymer phase. Truly miscible elastomer blends are rare, particularly in commercial use, the most notable exception probably being blends of poly(butadiene) with a high vinyl content and natural rubber or cis-1,4-poly(isoprene).

These factors may be summarized as:

- Polymer ratio
- Phase morphology
- Interfacial adhesion/crosslinking
- Distribution of filler between the elastomers
- Distribution of plasticizer between the elastomers
- Distribution of crosslinks between the elastomers

An attempt is made to illustrate these in Figure 1.1, which is a representation of a section through a model vulcanized blend of two elastomers. Reference is made to Figure 1.1 as each factor is considered further.

Polymer ratio, the relative amounts of each elastomer, is readily controlled and is often dictated by the end use in mind. The blend depicted in Figure 1.1 has a polymer volume ratio of about 3:1. Phase morphology has two aspects – type and size. Whilst type of phase morphology – at its simplest whether one phase is dispersed within another, as in Figure 1.1, or whether both are continuous – is largely governed by the polymer volume ratio, other factors such as the relative viscosity of the two elastomer phases can play a role.

Control of phase size is more complicated; obviously the conditions of preparation of the blend have a major influence. Here, only mixing of

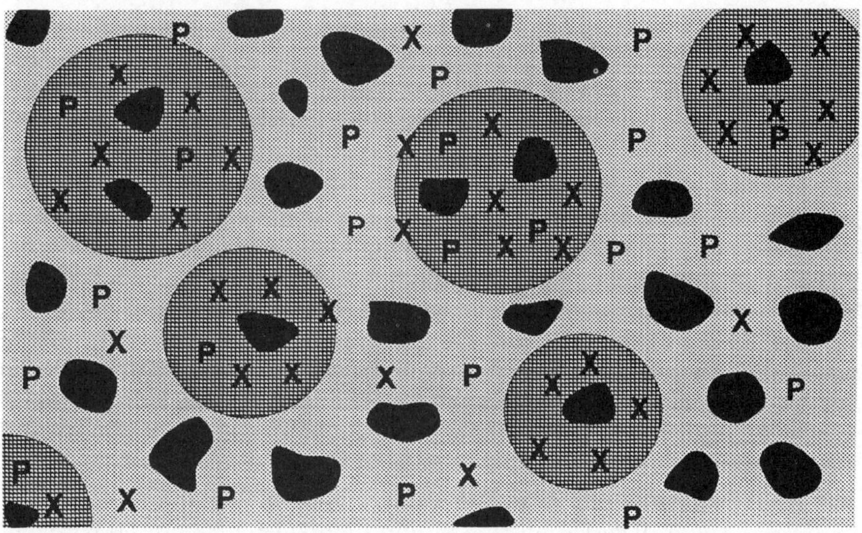

**Figure 1.1** A section of a model blend of two elastomers A and B, containing filler, plasticizer (**P**) and crosslinks (**X**).

elastomers is considered. In general, the higher the shear used in mixing, the smaller the phase size attained, although some elastomers undergo chain scission during mixing and there can be an optimum degree of mixing before phase size increases again. In practice, there are often constraints imposed on the extent of mixing which may be used – due either to economic considerations or such pragmatic considerations as an unacceptably high batch temperature.

Interfacial tension imposes a limit on the phase size which can be attained under a given shear regime: the lower the interfacial tension, the smaller the phase size. This is one reason that blends of the general purpose rubbers, natural rubber (NR, cis-1,4-poly(isoprene)), cis-1,4-poly(butadiene) (BR) and poly(butadiene-co-styrene) (SBR), present little difficulty. The elastomers are all non-polar and have rather similar solubility parameters (16.7, 17 and 17.5 $MPa^{1/2}$ respectively), and hence low interfacial tension. When there is a large difference in solubility parameters of two elastomers, there is high interfacial tension and phase sizes will be large – unless some means of reducing interfacial tension is employed. This is one of the difficulties facing blends of natural rubber and poly(acrylonitrile-co-butadiene) rubbers (NBR), which have solubility parameters ranging up to about 22 $MPa^{1/2}$.

Interfacial tension also plays a role in the control of interfacial adhesion and crosslinking between the two elastomers – high interfacial tension means little mixing of the two elastomers at the interface. If there is little mixing of the two elastomers at the interface, the opportunity for crosslinking between the two is reduced, and this can cause a weakness at the interface. Although there is no scale in Figure 1.1, the sharp phase boundary suggests little phase mixing at the interface and hence also relatively large phase sizes.

The remaining factors listed relate to distribution of entities between the polymer phases of the blend. Distribution of filler has been considered often in the past. In Figure 1.1, the filler appears to be essentially evenly distributed between the two phases. Distribution of the most commonly used filler, carbon black, is generally readily controllable by preparing well-mixed masterbatches of each elastomer containing the desired loadings of black. The generally good interaction of the elastomer with the surface of the carbon black, forming the so-called bound rubber, ensures that there is little transfer of filler between the two elastomers on crossblending of the masterbatches. Whilst a marked maldistribution of carbon black may cause difficulties in some blends, the ability to control filler distribution in this way can be put to good use – as will be seen in Chapter 7 dealing with a practical use of blends of NR and NBR and in the development of novel high damping blends of NR with epoxidized natural rubber (ENR) described in Chapters 10 to 13.

Distribution of plasticizer has not been considered so much in the past, perhaps because many compounds do not contain large quantities of plasticizer. However, when substantial quantities of plasticizer must or may be used, the distribution of this ingredient can affect properties – and can be used to advantage, as is evident in the development of the high damping NR/ENR blends described in Chapters 10 to 13. The idealized blend depicted in Figure 1.1 has a maldistribution of plasticizer in favour of the continuous phase (A).

The possibility of uneven distribution of crosslinks between the phases of a vulcanized blend has long been seen as a potential cause of poor physical properties, particularly within the context of blends of highly unsaturated elastomers, such as NR, with elastomers containing only a low level of unsaturation, such as ethylene-propylene-diene rubber (EPDM), as described in Chapter 14. This represents one source of a maldistribution of crosslinks – a substantial difference in the concentration of potential crosslink sites in the two elastomers, since the unsaturation is necessary for sulphur vulcanization. An alternative source is a difference in the concentration of the other reactants involved in the crosslinking reactions: sulphur, accelerators and species derived from these during the vulcanization process. Partition of these chemicals between the elastomers is to be expected when there is some mechanism for preferential solubility. The large difference in polarity and solubility parameters of NR and NBR provides a good example. The disperse phase (B) of the blend depicted in Figure 1.1 is more highly crosslinked than the continuous phase.

Systematic study of the distribution of crosslinks between the phases of vulcanized blends has only become possible recently with the development of a reliable technique for characterization of crosslink density in the individual phases of a blend. This is described in the following chapter. The work on overcoming the difficulties which arise in some blends described in subsequent chapters has relied to some extent or another on the ability to measure and control crosslink distribution and this book represents the most comprehensive account of the deployment of the technique for the benefit of materials development.

The factors described above provide ample opportunity for difficulties to arise in the practical use of elastomer blends, particularly when the elastomers differ fundamentally in character. This is almost always the case when attempts are made to combine desirable attributes of general purpose and special purpose or speciality elastomers. However, under favourable circumstances these factors also provide the means of overcoming practical difficulties and may even be exploited to produce blends which exhibit a new range of properties.

## 1.2 BLENDS CONSIDERED

Four blend systems are addressed here, each recognizing the needs of particular types of application:

- Blends of natural rubber and nitrile rubber (NR/NBR) for applications requiring a measure of resistance to swelling by oils or of damping combined with good physical properties.
- Blends of natural rubber and ethylene-propylene rubber (NR/EPM) for applications requiring very high resistance to attack by ozone.
- Blends of natural rubber and ethylene-propylene-diene rubber (NR/EPDM) for applications requiring a combination of good resistance to attack by ozone and good strength properties.
- Blends of natural rubber and epoxidized natural rubber (NR/ENR) for applications requiring a combination of high damping, good physical properties and low dependence of dynamic properties on temperature.

Whilst general problems have been addressed for each blend, at least one specific application has been targeted to focus the work. These have been prompted by known needs on the part of the rubber product manufacturing industry.

## 1.3 STRATEGY

The strategy adopted for each of the blend systems was to:

- confirm general principles behind a new approach to overcoming practical difficulties,
- identify technically and economically viable solutions to the difficulties,
- establish rules of use of the solutions where appropriate,
- demonstrate the feasibility of the solutions in a target application.

The overall thrust was to enable the rubber product manufacturing industry to use its own knowledge and resources to elaborate the solutions for particular problems within its market sector.

## 1.4 WIDER IMPLICATIONS

There has been considerable success in overcoming the problems posed by the blends considered here, and in realizing some of the opportunities to provide new alternatives for elastomers in a range of applications. Whilst these developments have been made within the context of essentially four specific blend systems, the solutions may be more widely applicable.

Some care must be exercised in generalizing the particular solutions found to the problem of obtaining even crosslink distribution in NR/

NBR blends. It is evident that a cure system which is ideal for blends with NBR at one level of acrylonitrile content may well be quite inappropriate for blends with NBR at a somewhat different acrylonitrile content. Nonetheless, the principle of identifying cure systems which will give near-even crosslink distribution should be applied to all blends of elastomers differing in polarity. Furthermore, one general rule has emerged – it appears that the very polar thiuram accelerators, TMTM and TMTD, should be avoided in NR/NBR blends.

The reactive mixing approach which proved so successful in NR/EPDM blends should also have wider applicability, and offers the prospect of a solution to difficulties with other blends of highly unsaturated elastomers and elastomers with low levels of unsaturation.

Again, the approach adopted to developing blends with a combination of high damping, good physical properties and a low dependence of properties on temperature could be applied to other blends of elastomers differing in polarity. However, the particular combination of NR with ENR has the distinct advantage that both elastomers are inherently high strength due to strain-induced crystallization.

# 2

# Measurement of crosslink density in vulcanized blends

*Andrew J. Tinker*
*Tun Abdul Razak Research Centre, MRPRA, UK*

## INTRODUCTION

Much of the work described in this book has been enabled or assisted by the ability to measure crosslink densities of the individual rubbers in a vulcanized blend. Indeed, for two of the blend systems, NR/NBR and NR/EPDM, the problems were to some extent defined by earlier measurements of crosslink densities. From these earlier studies, it was known that preferential distribution of crosslinks in favour of one of the elastomer components of NR/NBR blends, particularly the NBR, was likely to occur to an extent sufficient to explain the generally disappointing physical properties obtained [1, 2]. The failure of EPDM to vulcanize adequately in the presence of the highly unsaturated NR has been known for many years [3–5], but measurements of crosslink densities of blends of NR with chemically modified EPDM showed that good physical properties could be obtained with only modest increases in crosslinking of the EPDM [6].

The techniques which have been proposed for estimating crosslink densities in vulcanized blends have been reviewed recently [7]. Of these, only swollen-state NMR spectroscopy has been used by several independent workers and thus received independent verification [8, 9]. Swollen-state NMR spectroscopy has been used extensively in the work described here. A second method has been used also, so-called 'network visualization'. This technique involves viewing specially prepared ultra-

*Blends of Natural Rubber*. Edited by Andrew J. Tinker and Kevin P. Jones.
Published in 1998 by Chapman & Hall, London. ISBN 0 412 819406

thin sections of vulcanizates at high magnification by transmission electron microscopy (TEM). Whilst the technique is not ideally suited for estimating crosslink densities for vulcanizates with modest to high crosslink densities embracing levels normally found in practical materials, it can allow an insight into crosslinking between elastomers in a blend. This information is difficult to obtain by other means.

These two techniques are described here in some detail to provide a background for subsequent chapters which report results obtained by using them.

## 2.1 SWOLLEN-STATE NMR SPECTROSCOPY

Estimation of crosslink densities in the individual phases of a vulcanized rubber blend by swollen-state NMR spectroscopy is based on the straightforward observation that signals in NMR spectra of polymers are broader than those of simple molecules, and that signal width increases as the polymer is crosslinked (Figure 2.1). The basis of this observation is a progressive reduction in mobility. Spectra of vulcanizates cannot normally be obtained using a liquids NMR spectrometer, but swelling vulcanizates to equilibrium permits observation of spectra with sufficient resolution to identify signals from protons in different environments, although peaks are much broader than is normal for polymer solutions.

The effect of crosslink density on signal width is readily evident simply by comparison of spectra, but may be quantified through esti-

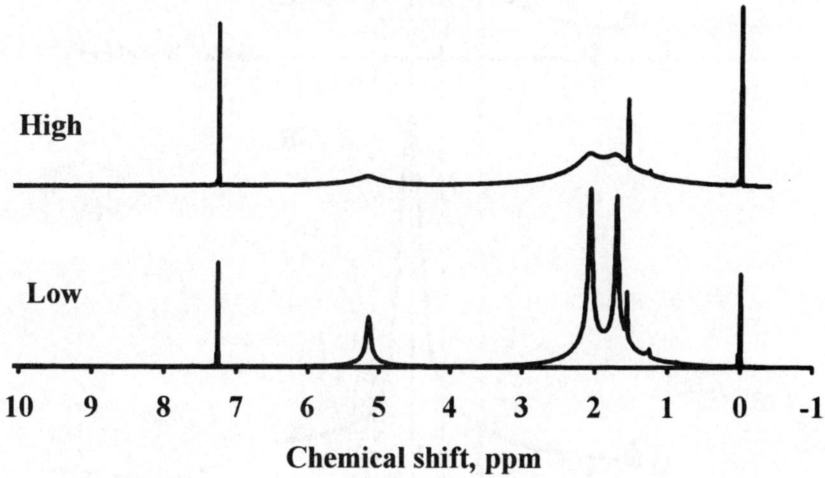


**Figure 2.1** Comparison of swollen-state $^{1}$H NMR spectra of gum vulcanizates cured with S/CBS to have high (3.0/0.8 phr) and low (1.0/0.2 phr) crosslink densities.

mation of signal width. The procedure adopted allows a simple method of deconvolution, which is necessary when $^1$H NMR spectra of blends are considered because of the overlap of signals from the individual rubbers in the blend. Thus, the width of a particular peak, generally the signal due to the olefinic proton for unsaturated elastomers, is given as the signal strength at a reference position on the side of the peak expressed as a percentage of the peak signal strength and designated H% [10]. Hence, from Figure 2.2:

$$\text{H\%} = 100b/a$$

The position of the reference position is selected on the basis of providing a good range of H% over the range of crosslink densities of interest. H% can be correlated with physical crosslink density, $n_{\text{phys}}$, [10, 11] as estimated by conventional procedures such as from the Mooney–Rivlin constant $C_1$ [12] or equilibrium swelling and application of the Flory–

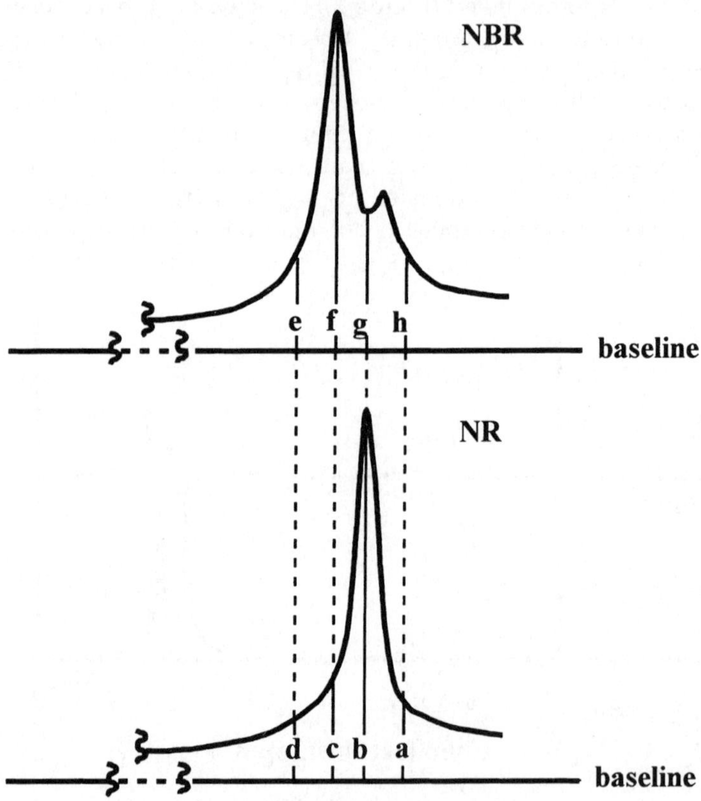


**Figure 2.2** Schematic representation of the olefinic signal in $^1$H NMR spectra of swollen gum vulcanizates of NR and NBR.

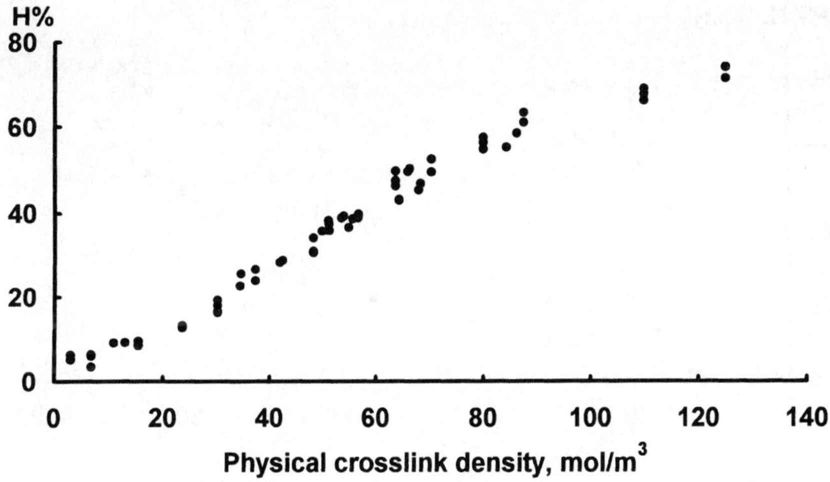


**Figure 2.3** Dependence of peak width, H%, on crosslink density for NR.

Rehner relationship [13, 14] (Figure 2.3). Correlations may also be made with indicators of crosslink density such as equilibrium swelling, rheometer torque rise or curative level used in vulcanization [10].

The simple deconvolution procedure involves definition of two correction terms at the peak and reference positions of the designated signal of the second elastomer. Again, these two terms are simply the signal strength at the position expressed as a percentage of the signal strength at the peak for the first elastomer. These two terms, P%H and R%H for the positions corresponding to the peak and reference positions for the second elastomer respectively, are given by reference to Figure 2.2 as:

$$\mathrm{P\%H} = 100c/a$$

and

$$\mathrm{R\%H} = 100d/a$$

Both terms may be correlated with H%, as shown in Figure 2.4.

Once these three correlations (H% with $n_{\text{phys}}$, P%H and R%H with H%) have been established for two elastomers, spectra of blends may be analysed to provide estimates of crosslink density in the two phases. H% is calculated for one of the elastomers without any allowance for a contribution to the signal by the second elastomer. Appropriate values for P%H and R%H may then be calculated and thence, from the peak signal strength, the magnitude of the corrections to be applied to the observed signal strengths at the peak and reference positions for the second elastomer. These corrections are subtracted from the observed signal strengths, and it is then possible to calculate the first estimate of

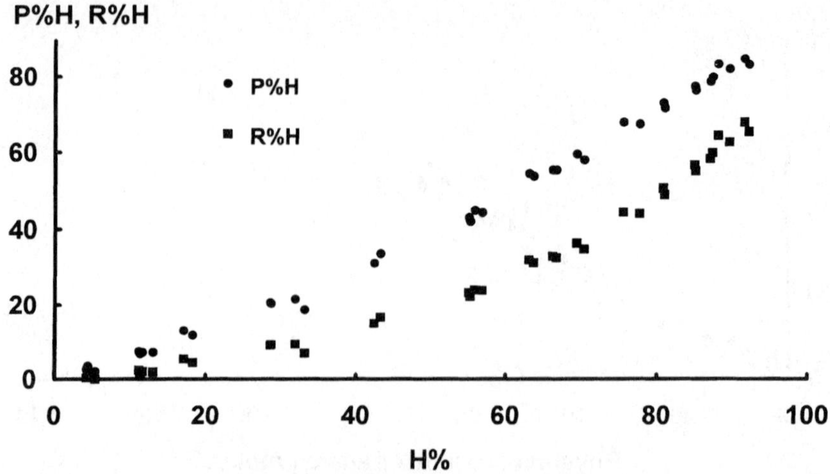


**Figure 2.4** Dependence of correction terms P%H and R%H on H% for NR.

H% for the second elastomer. The correction terms for the second elastomer and the magnitudes of the corrections to be applied to the observed signal strengths at the peak and reference positions for the first elastomer are calculated and applied. This represents the first cycle of an iterative process, which is repeated until changes in the estimates for H% become insignificant.

This process is both illustrated and proved in a simple demonstration [10]. Swollen samples of two different vulcanizates, NR and NBR for instance, may be introduced into one NMR tube and the spectrum acquired. This procedure was first performed using a 90 MHz continuous wave spectrometer, and the resulting composite spectrum is reproduced in Figure 2.5. The iterative process was applied to the spectrum, as shown in Table 2.1. The final estimates of H% for the two vulcanizates are the same regardless of which is chosen to start the calculation, the NR vulcanizate in the example quoted in Table 2.1. The estimates of H% for the two were in very good accord with estimates made separately, as noted at the foot of Table 2.1.

Physical crosslink densities may be obtained from the final estimates of H% by using calibration curves such as those in Figure 2.3.

The first reports [1, 10, 15–18] of crosslink densities in blends using this technique were based on spectra obtained with a 90 MHz continuous wave spectrometer. The technique was later transferred to a higher frequency 300 MHz FT spectrometer, with considerable benefits in terms of improved signal-to-noise ratio and resolution and reduced acquisition times. However, several important differences were noted [11], perhaps the most important being the observation that the position of peaks is

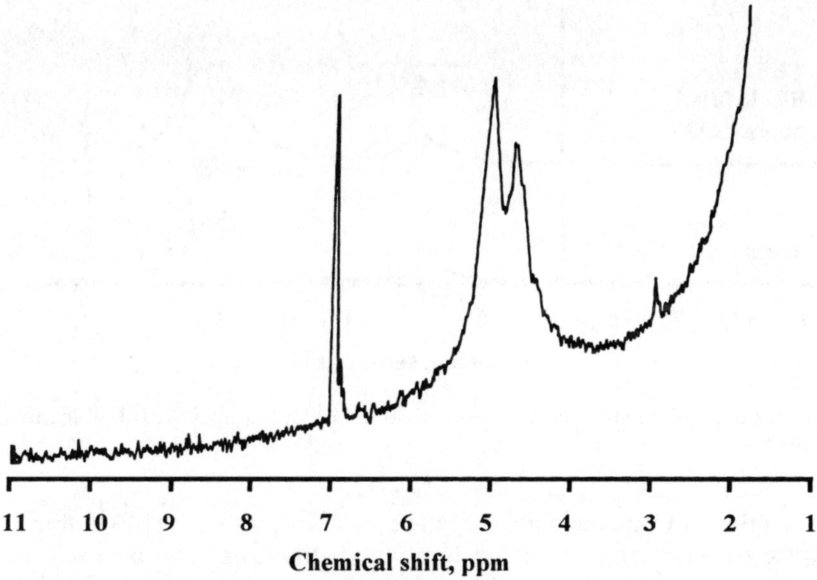


**Figure 2.5** 90 MHz $^{1}H$ NMR spectrum of slivers of swollen NR and NBR vulcanizates in the same NMR tube.

**Table 2.1** Data from spectra of samples of NR and NBR gum vulcanizates in a single tube

| *Iteration* | *H%* | | *Peak signal* | |
|---|---|---|---|---|
| | *NR* | *NBR* | *NR* | *NBR* |
| 0 | – | – | 0.556 | 0.665 |
| 1 | 59 | 24 | 0.492 | 0.386 |
| 2 | 56 | 30 | 0.466 | 0.431 |
| 3 | 55 | 32 | 0.456 | 0.449 |
| 4 | 54 | 33 | 0.451 | 0.457 |
| 5 | 54 | 33 | 0.449 | 0.460 |

Measurements on individual slivers in individual tubes: NR H% = 54, NBR H% = 35.3

dependent on crosslink density due to a difference in the local field within the swollen gel. Solvent within the gel also experiences this different field and thus the TMS added as a marker, for instance, has a sharp signal at 0 ppm and a smaller, broader signal at a position slightly downfield. The former is due to TMS in the free deuterochloroform outside the swollen vulcanizate, whilst the latter is due to TMS within

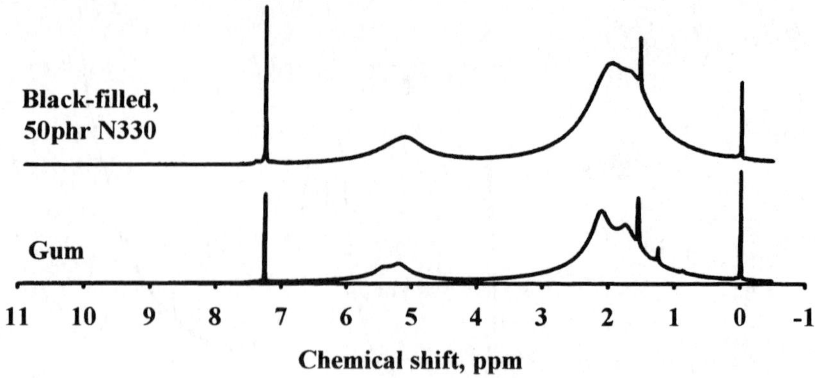


**Figure 2.6** Swollen-state NMR spectra of filled and gum NR/BR vulcanizate blends.

the swollen vulcanizate. Since both the elastomer and the TMS within the gel are experiencing the same local field, the latter can be used as a marker and signals do not shift relative to this marker with changing crosslink density.

Although carbon black filler causes broadening of peaks in NMR spectra of swollen vulcanizates (Figure 2.6), it is possible to investigate black-filled blends [11, 16, 19]. It should be noted that carbon black also affects peak positions, although again peaks do not shift relative to the secondary signal due to TMS in the swollen vulcanizate. Unlike crosslinking, increasing carbon black loadings cause a progressive upfield movement of signals. Signal width is dependent not only on carbon black loading but also on surface area and, to a lesser extent, structure [11]. Thus, it is necessary to develop calibration curves for each grade of carbon black and loading of interest. It is also highly advisable to use crossblends of well-mixed masterbatches, rather than blending elastomers and adding carbon black, in order to ensure that each phase of the blend contains the requisite level of black. Reliable independent estimation of crosslink density in black-filled vulcanizates is problematic; calibration curves of H% as a function of curative level have, therefore, often been used for such vulcanizates.

Despite the extensive overlap of signals in $^1$H NMR spectra of swollen, black-filled vulcanizates exemplified by Figure 2.6, it still possible to use the simple procedures for estimating H% for the components of a blend described above for spectra of gum vulcanizates. The demonstration has been repeated with black-filled vulcanizates, and again agreement between estimates of H% for NR and BR vulcanizates from a composite spectrum of the two slivers in a NMR tube are in good agreement with values obtained from spectra of separate samples (Figure 2.7).

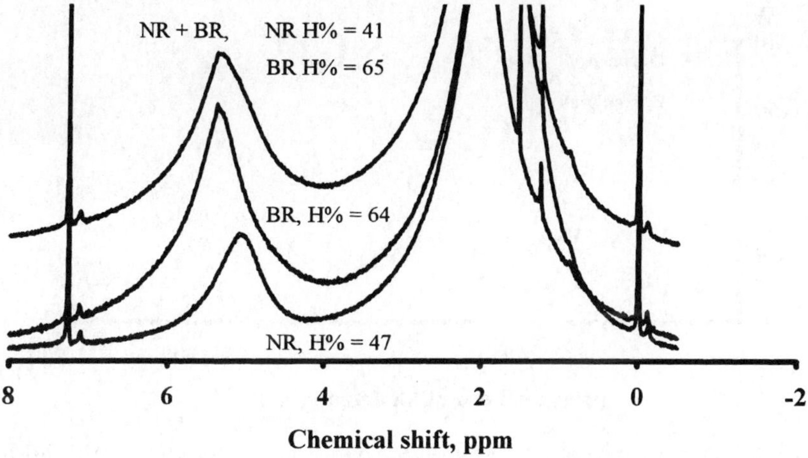


**Figure 2.7** NMR spectra of swollen black-filled vulcanizates (50 phr N330); NR alone, BR alone, NR and BR together.

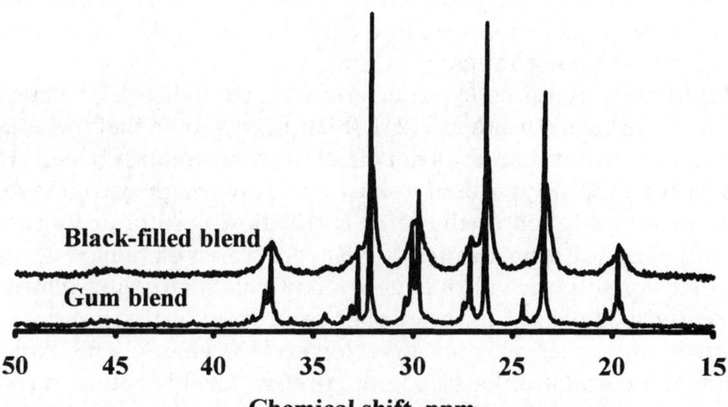


**Figure 2.8** Comparison of $^{13}$C NMR spectra of swollen gum and black-filled 60:40 NR:EPDM vulcanizates.

The use of a FT NMR spectrometer allowed investigation of blends with saturated elastomers. There is so much overlap of signals from aliphatic protons in $^{1}$H NMR spectra that it is not feasible to study saturated elastomers using such spectra, but $^{13}$C NMR spectra can be used due to good resolution of signals. Both gum and black-filled vulcanizates may be investigated (Figure 2.8). Good correlations of measures of peak width, either H% or the more conventional width at half-height ($W_{1/2}$), with $n_{\text{phys}}$ are observed [11, 20] (Figure 2.9). Good correspondence is seen

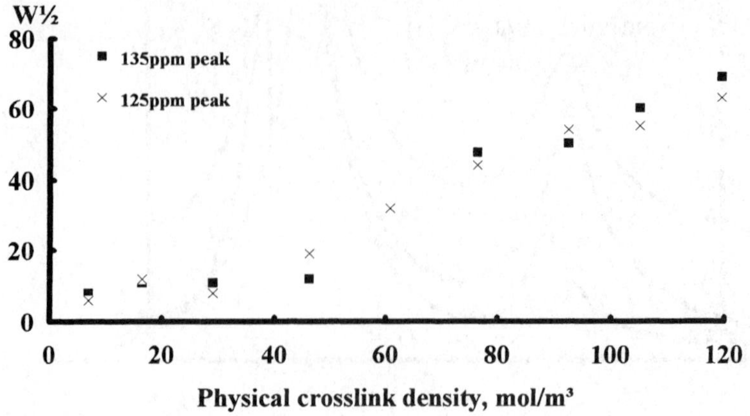


**Figure 2.9** Dependence of $^{13}C$ peak width on crosslink density for NR vulcanizates.

between estimates of $n_{phys}$ for NR in blends from $^1H$ and $^{13}C$ NMR spectra [11]. The only drawback in the use of $^{13}C$ NMR spectroscopy is the long accumulation times required for the acquisition of spectra with suitable signal-to-noise ratios.

Factors which might be expected to influence the swollen-state NMR technique have been considered [21]. It has been shown that the efficiency of the sulphur cure system does not affect the correlation between H% and $n_{phys}$ observed for any particular elastomer. Although peroxide vulcanizates were found to follow the same correlation as sulphur vulcanizates for cis-BR, peroxide vulcanizates of NR were observed to have lower H% values than for sulphur vulcanizates of equivalent crosslink density.

The molecular weight of the rubber prior to vulcanization influences H%: lower molecular weights give lower H% values. This can cause problems if a correlation of H% with curative level is being considered, as is often the case when black-filled vulcanizates are studied. However, physical crosslink density is also affected by the initial molecular weight of the elastomer, and a single relationship between H% and $n_{phys}$ is observed [21].

It may be anticipated that in swelling blends to equilibrium a phase may not be able to attain the degree of swelling normally associated with the prevailing crosslink density. Such swelling restriction may arise from either substantial differences in crosslink density in the individual phases or from a substantial difference in polymer–solvent interaction ($\chi$) values for the two elastomers. It has been shown that, whilst H% is influenced by the degree of swelling of the vulcanizate, this is of secondary importance [21]; the dependence on crosslink density is much stronger.

## 2.2 NETWORK VISUALIZATION MICROSCOPY

This technique is based on work by Shiibashi [22, 23]. Vulcanizate is swollen with styrene containing a low level of inhibitor, a peroxide and a small amount of phthalate plasticizer. Once the vulcanizate is swollen to equilibrium, the styrene is polymerized by raising the temperature. The resulting composite is sectioned, and the ultra-thin sections are stained with osmium tetroxide before viewing by TEM. At a suitably high magnification, a mesh structure is revealed (Figure 2.10). Although Shiibashi attributed the dark (stained) strands of the mesh to network chains, Cook *et al.* [24] have suggested that they represent bundles of network chains which form through phase separation as the concentra-

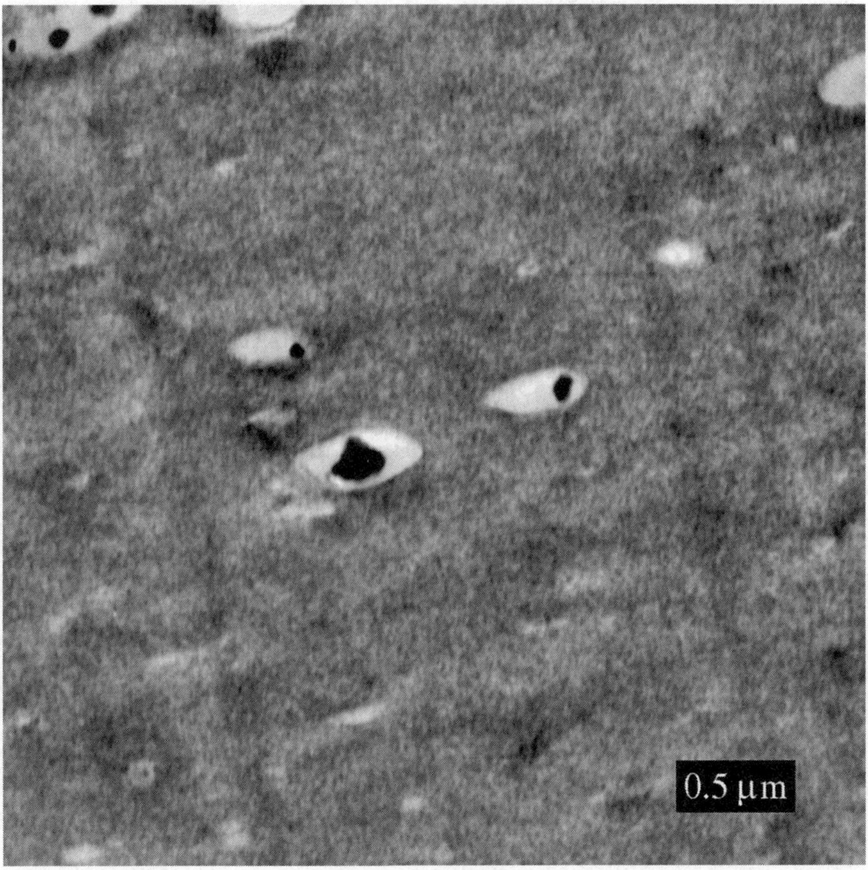


**Figure 2.10** TEM micrograph of a S/TMTM cured low crosslink density NR gum vulcanizate prepared for 'network visualization'.

tion of polystyrene rises during polymerization. Cook *et al.* observed a linear correlation between the mean mesh size, that is, the average size of the cells in the mesh structure, and molecular weight between physically effective crosslinks, $M_c$. $M_c$ is related to the physically effective crosslink density by:

$$n_{\text{phys}} = 1/2M_c$$

This observation forms the basis of measurement of crosslink density by the technique but, as noted in the Introduction, the estimates are prone to considerable error when crosslink densities are in the region of practical interest. However, the phase separation process provides useful information in a different way; regions of weakness in the vulcanizate are

**Figure 2.11** TEM micrograph of a S/MBS/MBTS cured 50:50 SMR L:BREON N41 gum vulcanizate prepared for 'network visualization'.

'exploited'. One example is the vicinity of particles of zinc oxide. The rubber is not bound to the zinc oxide, and a region of pure polystyrene forms around the particle as the surrounding vulcanizate is pushed away (Figure 2.10).

In vulcanized blends, this process can be used to obtain a qualitative assessment of the strength of the interface between the elastomers, and hence the extent of crosslinking between the two. If there is little crosslinking, the interface will be weak and regions of pure polystyrene will form as the interface is torn apart. An example of this is given in Figure 5.7 of Chapter 5. If crosslinking between the two elastomers is higher, the interface will be sufficiently strong to resist the forces of phase separation – either completely or at least until the polymerization process is more complete. In the former event, there will be no failure of the interface and the normal mesh structure only will be seen – as illustrated in Figure 2.11 whilst in the latter event, a limited failure will occur with the formation of only a thin region of polystyrene bridged by many strands or bundles of network chains. An example of the latter image is presented in Figure 5.8 of Chapter 5.

## 2.3 REFERENCES


1 Tinker, A.J. (1990) Crosslink distribution and interfacial adhesion in vulcanized blends of NR and NBR. *Rubber Chemistry and Technology*, **63**(4), 503–515.

2 Lewan, M.V. (1996) *Crosslink density distribution in natural rubber/nitrile rubber blends*. PhD Thesis, Loughborough University of Technology.

3 Corish, P.R. (1967) Fundamental studies of rubber blends. *Rubber Chemistry and Technology*, **40**(2), 324–340.

4 Rehner, J. Jr and Wei, P.E. (1969) Heterogeneity and crosslinking of elastomer blends. Investigations of chlorobutyl with highly unsaturated elastomers. *Rubber Chemistry and Technology*, **42**(4), 985–999.

5 Kerrutt, G. Blumel, H. and Weber H. (1969) Investigation of covulcanization. *Kautschuk Gummi Kunststoffe*, **22**(8), 413–418.

6 Brown, P.S. and Tinker, A.J. (1997) The effect of modifying EPDM on the crosslink distribution in NR/EPDM blends. Accepted for publication in *Journal of Natural Rubber Research*.

7 Tinker, A.J. (1995) Distribution of crosslinks in vulcanized blends. *Rubber Chemistry and Technology*, **68**(3), 461–480.

8 Shershnev, V.A., Shundrina, I.K., Yulovskaya, V.D. and Vasilenko, I.A. (1993) Regulating the degree of cross-linking in phases of elastomer blends on the basis of polyisoprene and polybutadiene. *Vysokomolekulyarnye Soedineniya, Series B*, **35**(10), 1708–1711; translated in *Polymer Science*, **35**(10), 1428–1431.

9 Van Duin, M., Krans, J.C.J. and Smedinga, J. (1993) Covulcanization and ozone resistance of NR/EPDM blends. *Kautschuk Gummi Kunststoffe*, **46**(6), 445–451.

10 Loadman, M.J.R. and Tinker, A.J. (1989) The application of swollen-state CW–$^{1}$H NMR spectroscopy to the estimation of the extent of crosslinking in vulcanized polymer blends. *Rubber Chemistry and Technology*, **62**(2), 234–245.

11 Brown, P.S., Loadman, M.J.R. and Tinker, A.J. (1992) Applications of FT–NMR to crosslink density determinations in natural rubber blend vulcanizates. *Rubber Chemistry and Technology*, **65**(4), 744–760.
12 Mullins, L. (1956) Determination of degree of crosslinking in natural rubber vulcanizate. Part I. *Journal of Polymer Science*, **19**, 225–236.
13 Flory, P.J. and Rehner, J. (1943) Statistical mechanics of cross-linked polymer networks. II Swelling. *Journal of Chemical Physics*, **11**(11), 521–526.
14 Flory, P.J. (1950) Statistical mechanics of swelling of network structures. *Journal of Chemical Physics*, **18**(1), 108–111.
15 Brown, P.S. and Tinker, A.J. (1990) Crosslink distribution in vulcanised blends of NR and EPDM. *Journal of Natural Rubber Research*, **5**(3), 157–157.
16 Brown, P.S. and Tinker, A.J. (1990) Estimation of the extent of crosslinking in the phases of vulcanized rubber blends by CW–$^1$H NMR spectroscopy. 3. Black-filled rubbers. *Communications IRC 90*, Paris, 48; also available as Publication 1320 from TARRC.
17 Brown, P.S. and Tinker, A.J. (1991) Distribution of crosslinks between the phases of vulcanised ENR-25/cis-BR blends. *Journal of Natural Rubber Research*, **6**(2), 87–95.
18 Brown, P.S. and Tinker, A.J. (1993) Distribution of crosslinks between the phases of vulcanised NR/cis-BR blends. *Journal of Natural Rubber Research*, **8**(1), 1–11.
19 Brown, P.S. and Tinker, A.J. (1995) The use of FT–NMR in the analysis of rubber blends: crosslink distribution in black filled blends of NR and cis-BR. *Kautschuk Gummi Kunststoffe*, **48**(9), 606–610.
20 Rooney, N.M., von Hellens, C.W., Tinker, A.J. and Brown, P.S. (1993) EPDM-phase crosslink density in blends of natural rubber and EPDM. Paper No. 128 presented at a meeting of the Rubber Division, American Chemical Society, Orlando, Florida, October 26–29; abstract (1994) in *Rubber Chemistry and Technology*, **67**(1), 198.
21 Brown, P.S. and Tinker, A.J. (1990) Factors affecting the NMR technique for estimation of crosslink density in rubber blends. *Journal of Natural Rubber Research*, **5**(4), 286–295.
22 Shiibashi, T. (1987) Gel structure characterisation of NR and IR and direct observation of individual polymer molecules by electron microscopy. *Nippon Gomu Kyokaishi*, **60**(6), 298; translated (1987) in *International Polymer Science and Technology*, **14**(12), T/33–T/39.
23 Shiibashi, T., Hirose, K. and Tagata, N. (1989) Direct observation of individual polymer molecules by electron microscopy. *Kobunshi Ronbunshu*, **46**(8), 473–479.
24 Cook, S., Cudby, P.E.F. and Tinker, A.J. (1992) Direct observation of crosslink distribution in vulcanized blends through visualizing the networks. Paper No. 3 presented at a meeting of the Rubber Division, American Chemical Society, Nashville, Tennessee, November 3–6; also available as Publication 1446 from TARRC.

# 3

# Characterization of vulcanized blends by microscopy

*Paul Cudby*
*Tun Abdul Razak Research Centre, MRPRA, UK*

## 3.1 INTRODUCTION

The primary aims of this chapter are to explain how the images used throughout this book have been formed and how they can be interpreted. In terms of the scope of the material described, good microscopical results can only be obtained by enhancing the contrast between the phases in a blend where very little exists. The following is limited to presenting the techniques applied to obtaining such results and is not intended to be a full and thorough theoretical description of the techniques employed. Other sources should be sought if this is required [1].

It was initially anticipated that most of the required results could be acquired by light microscopy (LM) with some of the more high resolution work being carried out using transmission electron microscopy (TEM). However, it rapidly became obvious that these techniques were, for reasons to be described, inadequate for routine observations of the full variety of materials being developed. Fortuitously, several new techniques in the field of scanning electron microscopy (SEM) were under development at MRPRA at the time and could therefore be used where necessary. This included a device described by Hitachi Scientific Instruments for converting a scanning electron microscope to operate as a scanning transmission electron microscope (STEM). This device was built at MRPRA and was subsequently modified to improve its performance [2].

*Blends of Natural Rubber*. Edited by Andrew J. Tinker and Kevin P. Jones.
Published in 1998 by Chapman & Hall, London. ISBN 0 412 819406

Two terms used throughout this contribution require to be defined as follows:

Blend morphology: the **overall** form and arrangement of the phases in a blend.

Phase structure: the arrangement of the **individual** components within a single phase.

All of the work described was performed on thin or ultra-thin sections of blend vulcanizates cut using cryo-ultramicrotomes. This has generally proved to be the most time-consuming aspect of microscopy and, as will be outlined in the ensuing text, has often been instrumental in deciding which is the appropriate technique for examining a material.

## 3.2 LIGHT MICROSCOPY

Traditionally, light microscopy (LM) has been the starting point for any microscopical examination of new samples. Transmitted images are formed by looking at thin sections taken from the bulk of the sample, effectively giving a cross-sectional image. The popularity of this technique has lain largely in its ease of use. In the case of elastomer blends, however, there tends to be very little contrast between the phases in transmitted common light microscopy. For this reason a special type of LM has been used, namely phase contrast LM [3, 4], which exploits differences in the refractive indices between the two (or more) phases to produce contrast. This technique can produce reliable micrographs which give considerable information about blend morphology for a number of materials. In Figure 3.1, a micrograph of an NR/NBR blend, natural rubber can be determined as the darker phase from a knowledge of the refractive indices of the individual components and an observation of the morphology. The blend morphology can be observed with ease and examined for flow orientation, phase sizes and large variations thereof, and degree of co-continuity of the phases.

However, available resolution is a fundamental limitation to any LM technique. Resolution is defined as the minimum distance between two object features at which they can still be seen as two features [5]. The resolution of any LM technique cannot be better than 0.25 μm as defined by the Abbé or diffraction theory of imaging [6] as:

$$d = \frac{\lambda}{\mathrm{k}NA}$$

where $d$ = resolution
$\lambda$ = wavelength of light used
k = a constant
$NA$ = Numerical Aperture.

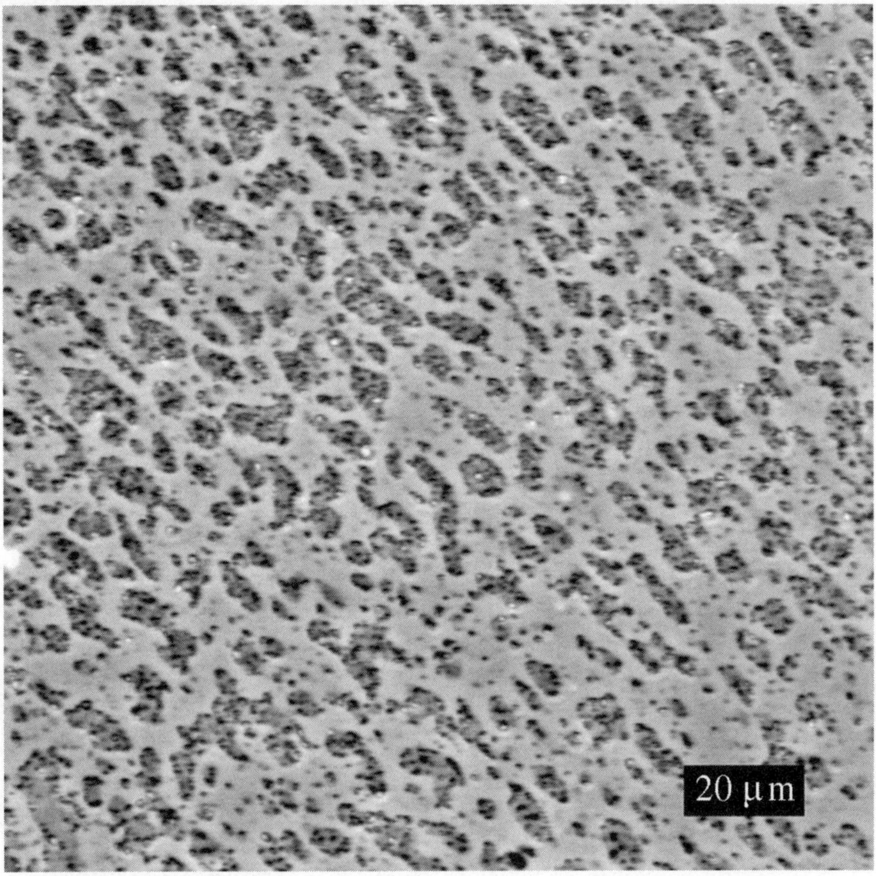


**Figure 3.1** Phase contrast micrograph of a 25:75 NR:NBR blend.

In practical terms, this means that no more useful information can be extracted from a light microscope beyond a magnification of about 1000×. However, with the rapid improvements in blend technology which have been covered in this book, phase sizes were being reduced to levels at which little useful information could be extracted using phase contrast LM. See for example Figure 3.2 which shows a blend of NR and EPDM with a calcium carbonate filler. Very careful observation may suggest the filler is preferential for the darker, NR phase, although such a conclusion lacks certainty. Furthermore, no information regarding phase structure or blend morphology can be determined from this micrograph. This blend is also considered subsequently during discussion of the SEM technique.

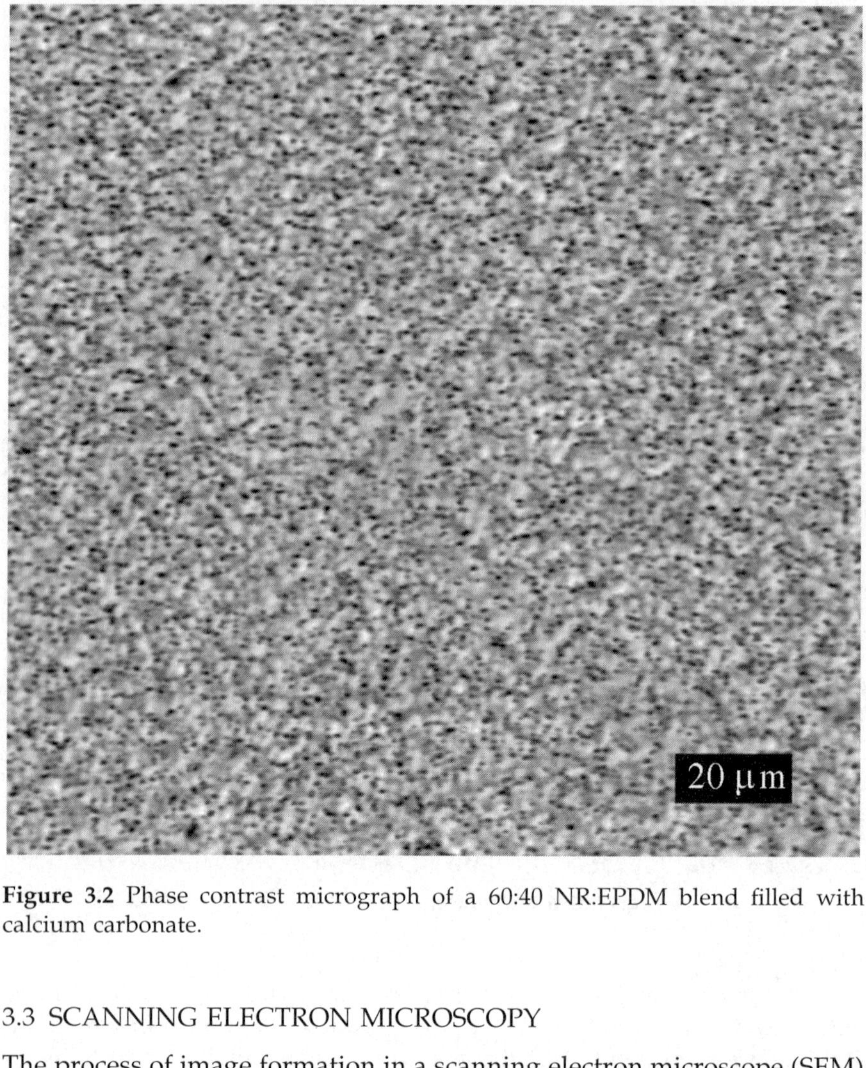

**Figure 3.2** Phase contrast micrograph of a 60:40 NR:EPDM blend filled with calcium carbonate.

## 3.3 SCANNING ELECTRON MICROSCOPY

The process of image formation in a scanning electron microscope (SEM) can be described in terms of a focused electron probe being scanned across the specimen surface, under high vacuum, while synchronized to the raster scan on a TV type display. Where the electron beam strikes the surface a number of different types of interaction take place including the emission of electrons. The image is formed point by point in accordance with the strength of whichever of the signals is measured from the individual scanned points as this modulates the strength of the corresponding point on the display.

This discussion is limited to only two of the various interactions which take place, namely those involving secondary electrons and backscattered electrons. In simplified terms secondary electrons are low energy electrons which have been inelastically scattered by atomic electrons in the sample, and backscattered electrons are high energy electrons which have been elastically scattered by atomic nuclei. Secondary electrons are largely responsible for contrast dependent on variations in the surface topography and backscattered electrons are largely responsible for contrast dependent on variations in elemental composition. This could be explained as: the greater the slope, the stronger the secondary electron signal and the higher the atomic number, the stronger the backscattered electron signal.

An ideal specimen could therefore be conceived to be similar to that shown in Figure 3.3, wherein an object made of a heavy metal, such as iron, protrudes significantly from the surface of a carbon substrate, thus maximizing both secondary and backscattered electron signals. Unfortunately, in the real world of polymer blends this idealized situation does not apply, with the reality being as depicted in Figure 3.4. The elastomers considered in the scope of this work consist of very similar elements, all of which have low atomic numbers and therefore give rise to little contrast due to the backscattered electron signals. Furthermore, blends are examined as sections, so there is very little variation in topography and hence very little contrast due to the secondary electron signal. Consequently, if examined directly, most sections of elastomers appear relatively featureless by SEM. Therefore, artificial techniques have to be introduced for increasing the contrast in the SEM. Two such techniques have been used extensively: chemical staining and chemical etching. These are described below.

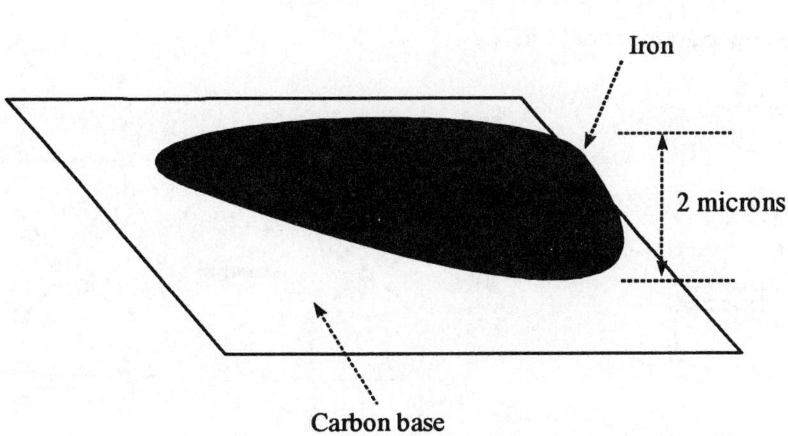


**Figure 3.3** SEM – idealized sample.

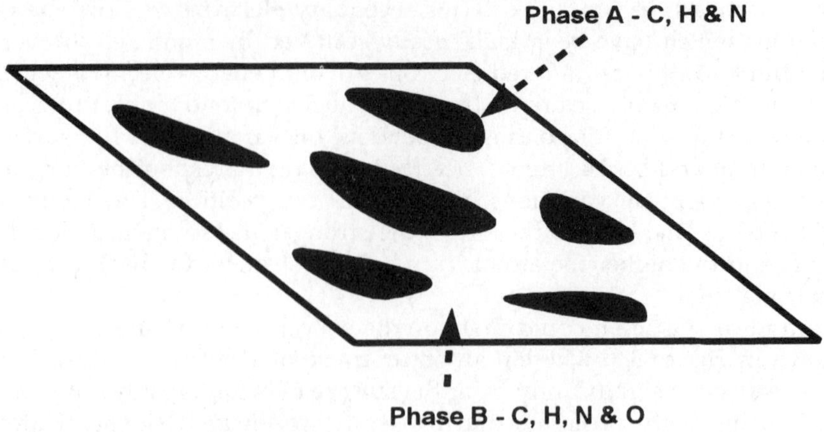


**Figure 3.4** SEM – phase morphology.

## 3.4 CHEMICAL STAINING

Although the elastomers used here are elementally similar, they are chemically quite different. It is therefore possible to preferentially react one phase with a chemical containing a heavy atom, thus producing elemental contrast. This is referred to as differential chemical staining. The most useful such stain with this work has been osmium tetroxide which reacts with unsaturated carbon–carbon bonds [7] as shown in Figure 3.5. Since the degree of unsaturation in different polymers is known, it is possible to predict which will be stained to the greater degree.

**Proposed mechanism:**

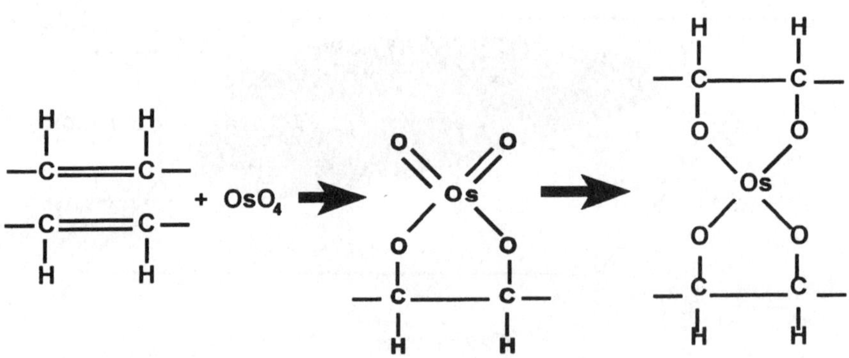


**Figure 3.5** Differential chemical staining with osmium tetroxide.

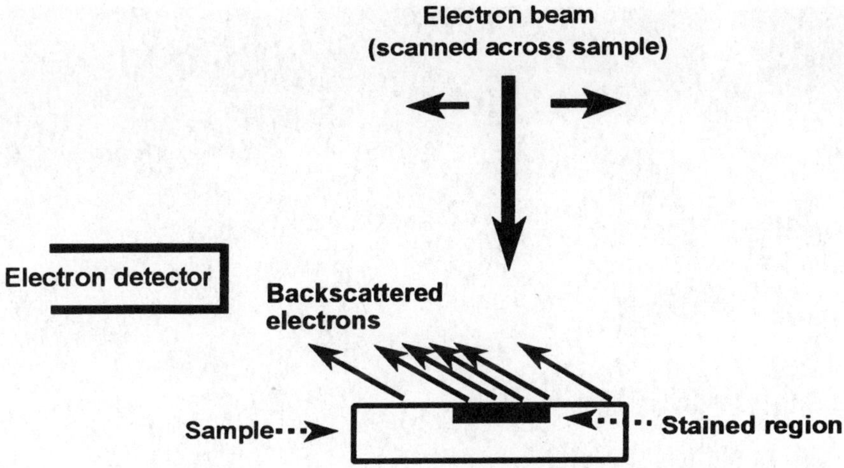


**Figure 3.6** Increasing contrast in the SEM.

The effect of staining is illustrated in Figure 3.6. Where a region has been stained by osmium tetroxide it will yield a greater proportion of backscattered electrons than a region which contains less, or no, osmium tetroxide. Therefore, the stained region will appear brighter. This yields the kinds of result shown in Figures 3.7 and 3.8 in which a blend of NR/NBR can be observed at a much higher resolution than is possible with the light microscope. The NR contains a greater degree of unsaturation and is consequently stained to a greater degree and appears brighter than the NBR phase. Measurements of phase size and observation of blend morphology can be accomplished easily from such micrographs.

## 3.5 CHEMICAL ETCHING

The alternative technique is differential chemical etching. For a predetermined time a chemical is introduced to the surface and chemically etches out one of the phases. This technique is less easy to accomplish and more care needs to be taken in interpretation. Figure 3.9 shows a section viewed at right angles to the viewing direction used in the SEM. From this diagram it is clear that overhangs of the unetched phase can give misleading information as to the sizes and shapes of the etched phase simply by obscuring the field of view of the observer from the space under such an overhang. In a similar manner, the observer may not easily recognize the blend ratio as being correct.

Some success has been achieved in observing NR/ENR blends such as that shown in Figure 3.10. This specimen has been etched with phosphotungstic acid which has etched away the NR phase whilst

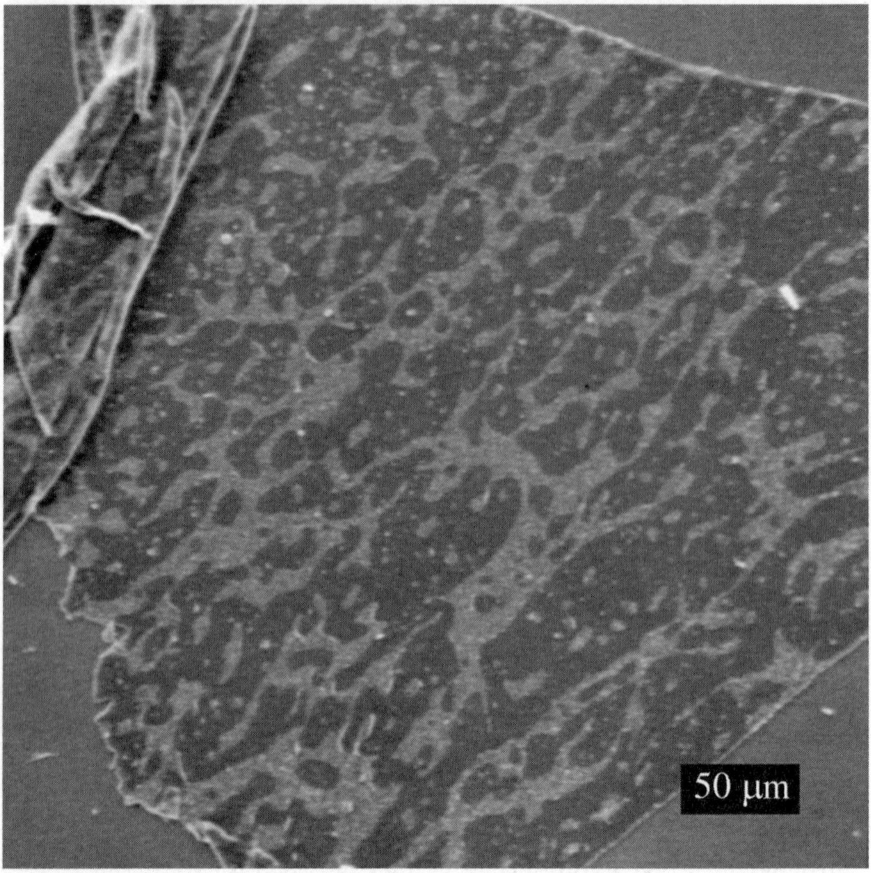


**Figure 3.7** SEM micrograph of a 30:70 NR:NDR blend stained with osmium tetroxide.

leaving the ENR-25 phase intact. The final appearance gives a better idea of the three-dimensionality than simple staining would, with the ENR-25 phase taking on the appearance of a sponge.

In summary, SEM can be described as a relatively fast technique for observing blend morphology at medium resolution. Nevertheless, the technique is not suitable for observing the internal phase structure of individual phases. To achieve this, sections must be viewed in transmission, as with LM.

## 3.6 TRANSMISSION ELECTRON MICROSCOPY

Transmission electron microscopy (TEM) provides the highest resolution, hence the sharpest images, of the conventional electron microscope

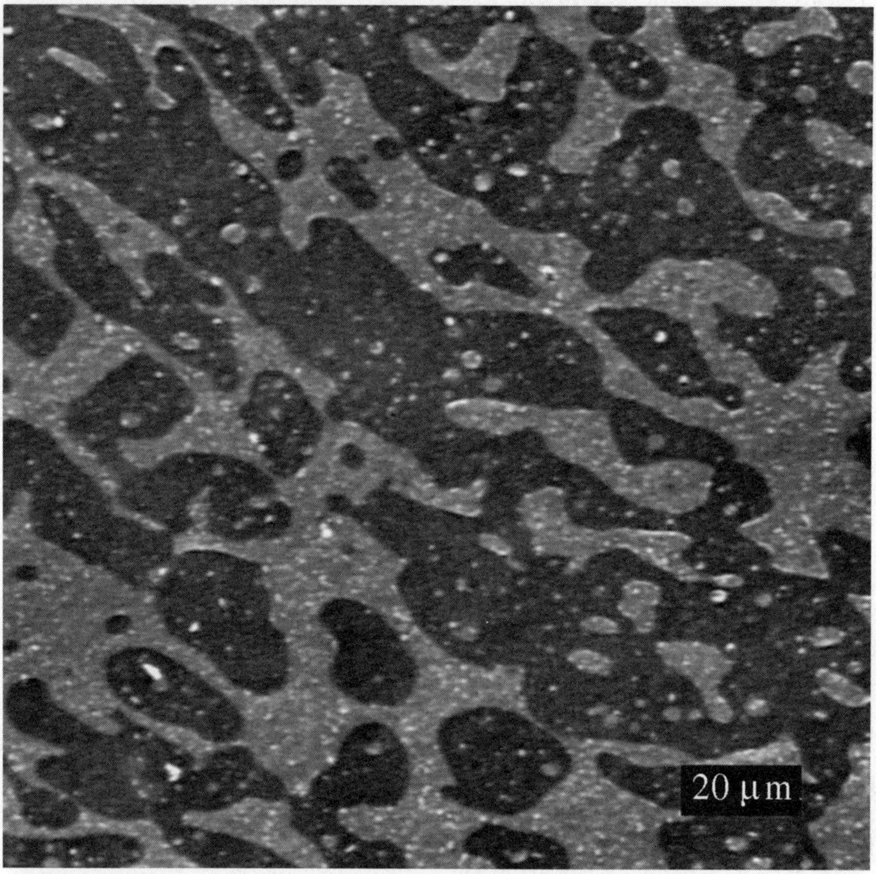


**Figure 3.8** SEM micrograph of a 30:70 NR:NBR blend stained with osmium tetroxide.

techniques. With a suitable sample, 0.34 nm can be achieved with the equipment available to the author. Therefore, this technique should give more visual information about blend morphology and phase structure than the other techniques available.

Images are formed in a TEM by focusing an electron beam onto an ultra-thin section (typically 150 nm thick or less) of the specimen. Electrons travel through the specimen, either deflected or undeflected, and are focused and enlarged by a series of electromagnetic lenses onto either a fluorescent viewing screen or a piece of electron-sensitive film. Contrast is dependent on a number of factors, but for this type of imaging the major factor is the degree to which different phases scatter electrons. However, as with SEM there are insufficient elemental differences between the materials used here to produce much contrast. It has therefore

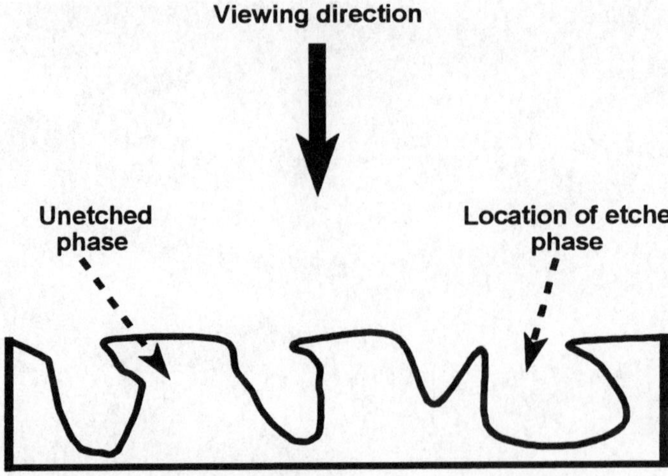


**Figure 3.9** SEM examination of an etched surface.

been necessary to stain materials, usually with osmium tetroxide, to produce the required results. However, as depicted in Figure 3.11, the mechanism for achieving this contrast is quite different. The beam is focused on to the region of interest, rather than scanned across it. A phase that has been preferentially stained with osmium tetroxide will scatter electrons further on average than an unstained region. With the use of a contrast limiting aperture, these scattered electrons are prevented from travelling further down the microscope and consequently the stained area appears to be dark rather than light as in the case of SEM.

Three examples are given of the type of micrograph obtained by TEM during the course of this work. Figure 3.12 shows an unfilled NR/EPDM blend. In this case, as in the stained phase, the NR appears dark. A number of key features now become apparent which could not have been observed with the resolution available using LM. The first of these is the phase structure of each phase wherein a captured micro-phase of each elastomer within the other phase can be seen; that is, small regions of NR are clearly present within the EPDM phase and vice versa. Secondly, the overall co-continuity of the morphology can be observed and in particular the way in which the EPDM phase passes through the section. By extrapolating on a mental level it is possible to visualize the sample from which the section was taken as a block of NR with interconnecting EPDM "tunnels" running throughout.

Figures 3.13 and 3.14 illustrate the calcium carbonate filled NR/EPDM blend first shown in Figure 3.2 as imaged by phase contrast light microscopy. In common with Figure 3.12 the overall morphology can now

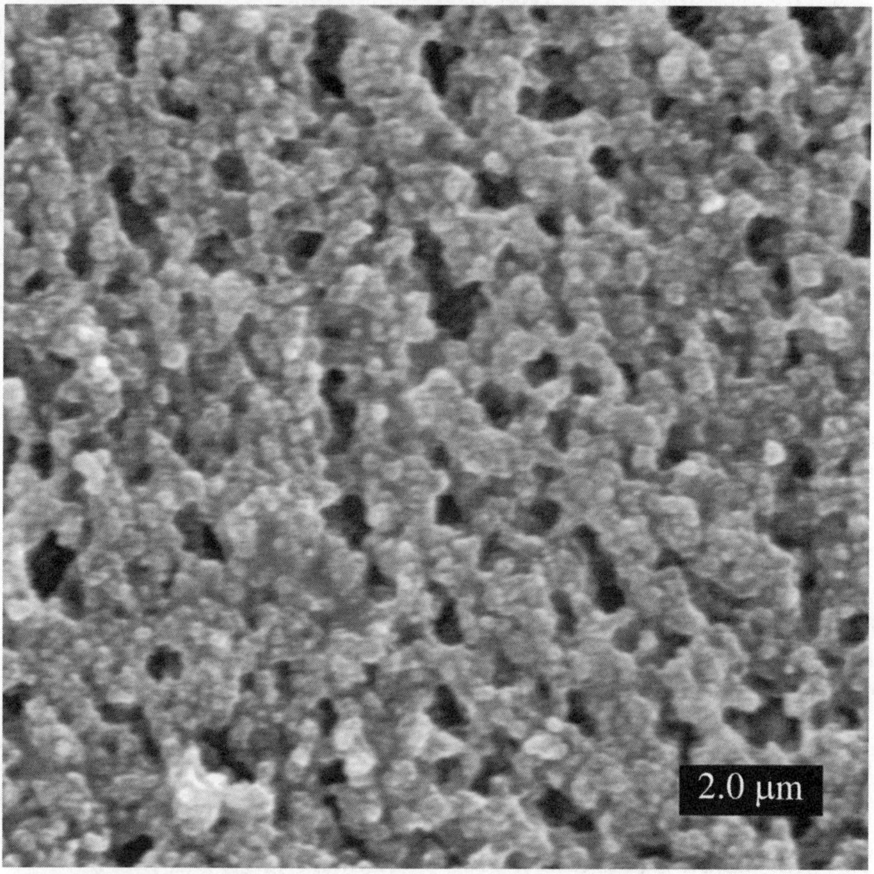


**Figure 3.10** SEM micrograph of a 50:50 NR:ENR-25 blend masterbatch etched and stained with phosphotungstic acid.

be seen far more clearly. By working at higher magnification it was possible to show that the filler appeared to be wholly located within the NR phase. Even in instances where the filler appeared to be within the EPDM phase, a close examination reveals that the particle(s) in question actually lie within an NR captured micro-phase within the EPDM phase.

There are some severe limitations with TEM, however. The most significant of these is that it is an extremely time-consuming technique. It requires ultra-thin sections, less than 150 nm thick, and preparation of sections of this type can take from two hours to several days depending upon the nature of the sample. For instance, if a difficult sample such as a highly black-filled specimen is encountered, with the number of variables involved in cryo-ultramicrotomy it can take several hours to determine the best conditions for obtaining sections which are thin enough

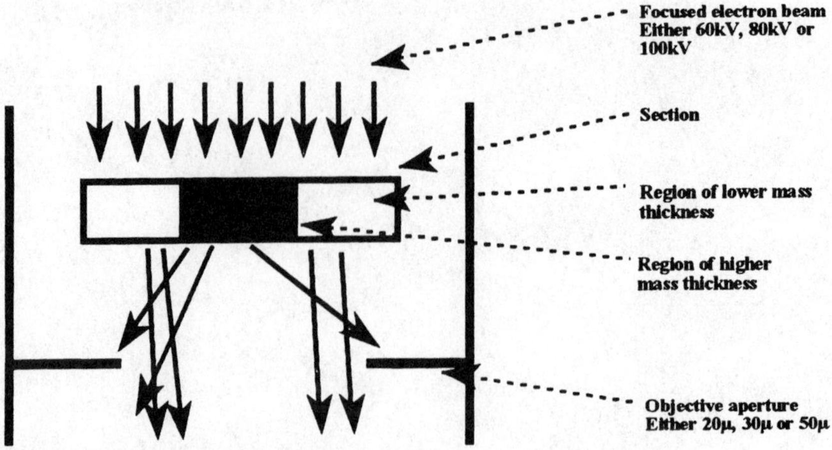


**Figure 3.11** Contrast formation in the TEM.

to be useful. In addition, the TEM cannot handle instant film, so negatives obtained in this manner have to be developed and printed separately, and under some circumstances it can be difficult to obtain low magnification micrographs. Ultimately, although vital for high resolution work, TEM has not proved practical for the routine assessment of phase morphology of large numbers of samples. Another technique is therefore needed to fill the apparent gap in transmitted techniques between phase contrast LM and TEM.

## 3.7 SEM-BASED SCANNING TRANSMISSION ELECTRON MICROSCOPY

SEM-based scanning transmission electron microscopy (STEM) is not a new idea, although TEM-based STEM is far more common. It could be described as the "missing link" between TEM and LM. The technique described here is based on a simple mount constructed at MRPRA, the design of which arose from communications with Hitachi Scientific Instruments, although modifications have been made since.

The theory behind the technique is somewhat of a hybrid between SEM and TEM imaging and can be best described via Figure 3.15. A brief examination may suggest that the mount is a form of electron "mirror", although a more detailed discussion of imaging theory shows this not to be the case. The STEM mount consists of several copper components. The top part of the mount consists of a hollow tube of 2.7 mm internal diameter incorporating a locking cylinder in which the prepared specimen, placed on a standard TEM examination grid, is held. This is positioned over a polished angled plate which, when placed in the standard spec-

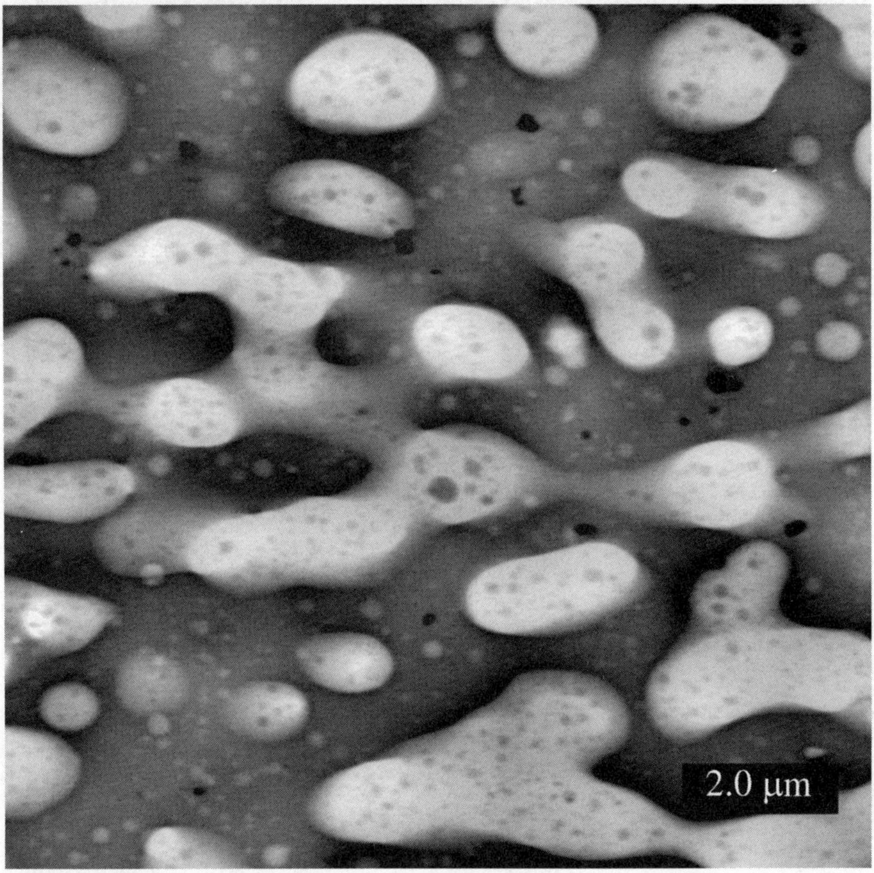


**Figure 3.12** TEM micrograph of a 60:40 NR:EPDM blend stained with osmium tetroxide.

imen stub holder in the SEM specimen chamber, is directed toward the secondary electron detector.

The principle is thus: the focused electron beam is scanned across the section and electrons strike the sample. The thickness of the section usually precludes any signal due to transmitted secondary electrons. The important portion of the signal is therefore due to forward-scattered as opposed to backscattered electrons which produce an image determined by mass and thickness of the regions being examined, as in the case of TEM. After transmission, these electrons strike the angled plate producing a secondary electron signal which is detected by the secondary electron detector.

STEM has several advantages over conventional TEM, SEM and LM. The added versatility imparted to the SEM by virtue of its digital elec-

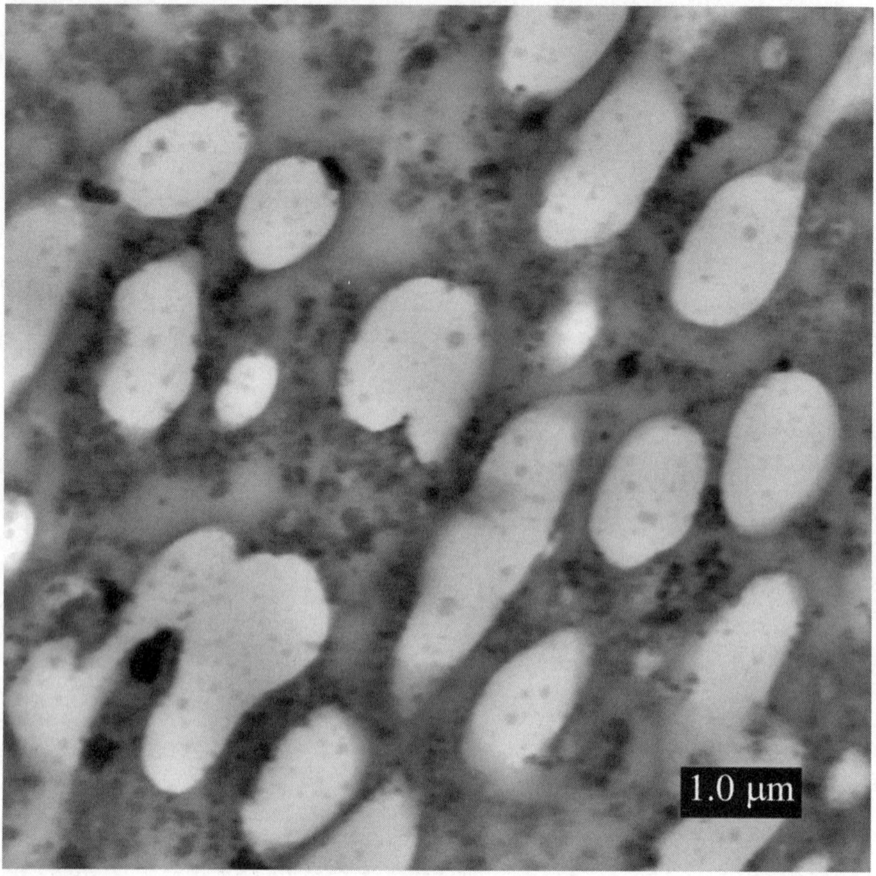


**Figure 3.13** TEM micrograph of a 60:40 NR:EPDM blend, filled with calcium carbonate, stained with osmium tetroxide.

tronics means that it is capable of imaging thicker sections than the TEM, although resolution is obviously improved with thinner sections, and since it can use Polaroid instant film (which the TEM cannot) results can be obtained more quickly. Furthermore, resolution is substantially improved over that obtained by LM. Since this technique is used for transmission imaging it provides information about both blend morphology and phase structure. When all of these time savings are combined, the value of SEM-based STEM as a routine transmission technique becomes clear.

The scope of STEM is illustrated firstly in Figure 3.16, a micrograph of an NR/NBR blend in which the NR is the darker phase. The most important features in this micrograph are the very light regions which have occurred at several points around the phase interfaces. These regions are

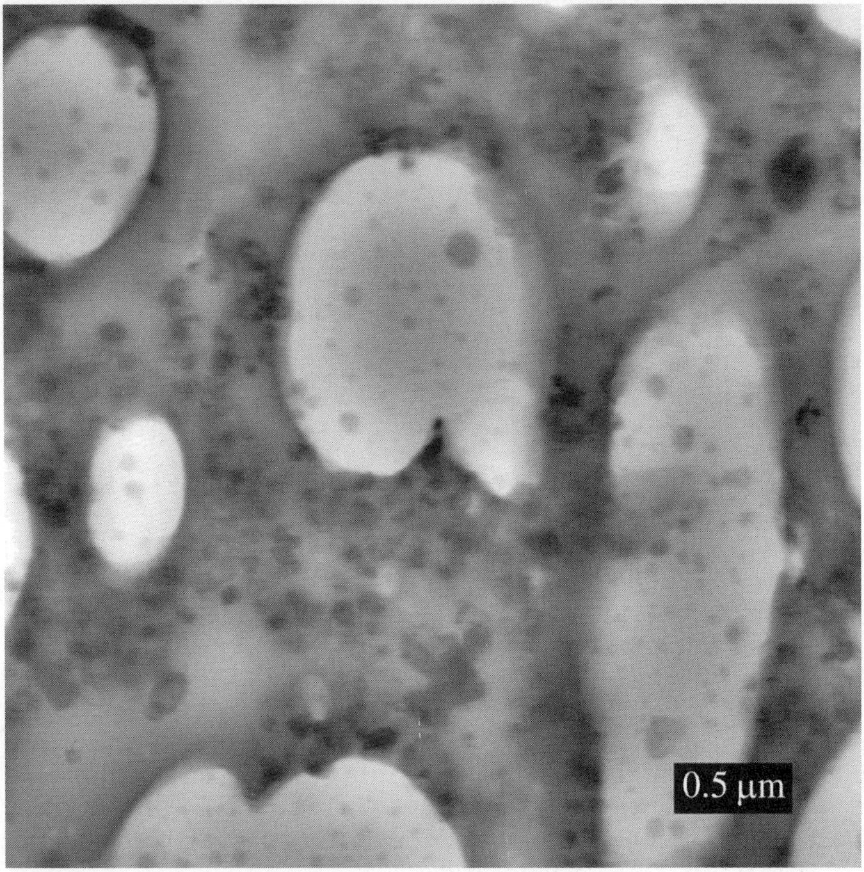


**Figure 3.14** TEM micrograph of a 60:40 NR:EPDM blend, filled with calcium carbonate, stained with osmium tetroxide.

failure points at the interface where the two phases have separated. Because there is no material in the way of the beam, it can pass through unobstructed, thus making these regions appear to be very light. Holes around the captured micro-phase can also be observed where the action of sectioning has literally pushed such a region out of its position within the matrix. The removed pieces can often be seen next to the hole, in some cases still attached to some point on the interface. Weak interfacial adhesion within NR/NBR blends has been an important theme running throughout this work: results like these have been vital to the work performed in this field (Chapters 5 to 8).

Figure 3.17, showing the same unfilled NR/EPDM as observed by TEM in Figure 3.12, has been included to provide a direct comparison, and to demonstrate the validity of the STEM technique. Whilst the res-

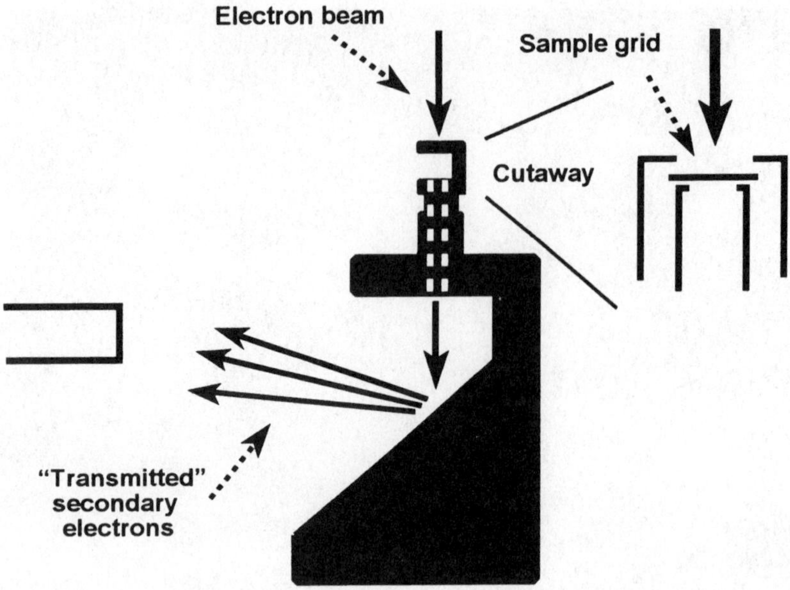


**Figure 3.15** Cutaway of the STEM mount.

olution is obviously not as high as observed by TEM, all of the important information about blend morphology and phase structure (and microstructure) is included in the micrograph. The advantage is that this micrograph was obtained far more quickly than would have been possible via TEM.

## 3.8 CONCLUSION

For many years conventional and phase contrast light microscopy have been the starting point for routine examination of blends for the assessment of overall phase morphology at low magnification. However, advances in blend technology have reduced phase sizes to such an extent that the limited resolution afforded by light microscopy has become insufficient to make useful observations. The alternatives are the various forms of electron microscopy, each of which has advantages and limitations. The three techniques used here were scanning electron microscopy, transmission electron microscopy and SEM-based scanning transmission electron microscopy. The use of these three types of electron microscopy has contributed far more morphological information than could have been obtained from a reliance on light microscopy alone, or indeed any one single microscopy technique.

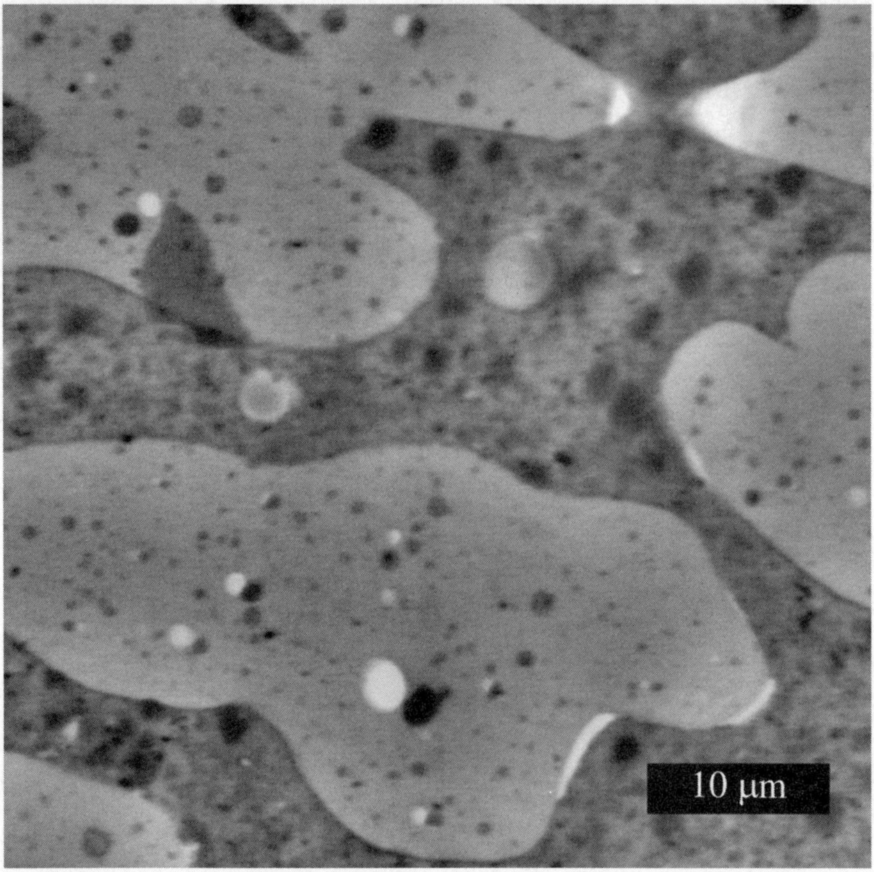


**Figure 3.16** STEM micrograph of a 30:70 NR:NBR blend stained with osmium tetroxide.

The characteristics of the types of microscopy used are summarized below:

- LM: good starting point for obtaining blend morphological information.
- SEM: higher resolution than LM: useful for looking at overall blend morphology, but not for providing easy interpretation of phase structural details.
- TEM: provides very high resolution images which contain a large amount of information: can be time consuming but remain essential for some aspects of the research including the swollen network technique described in Chapter 2.

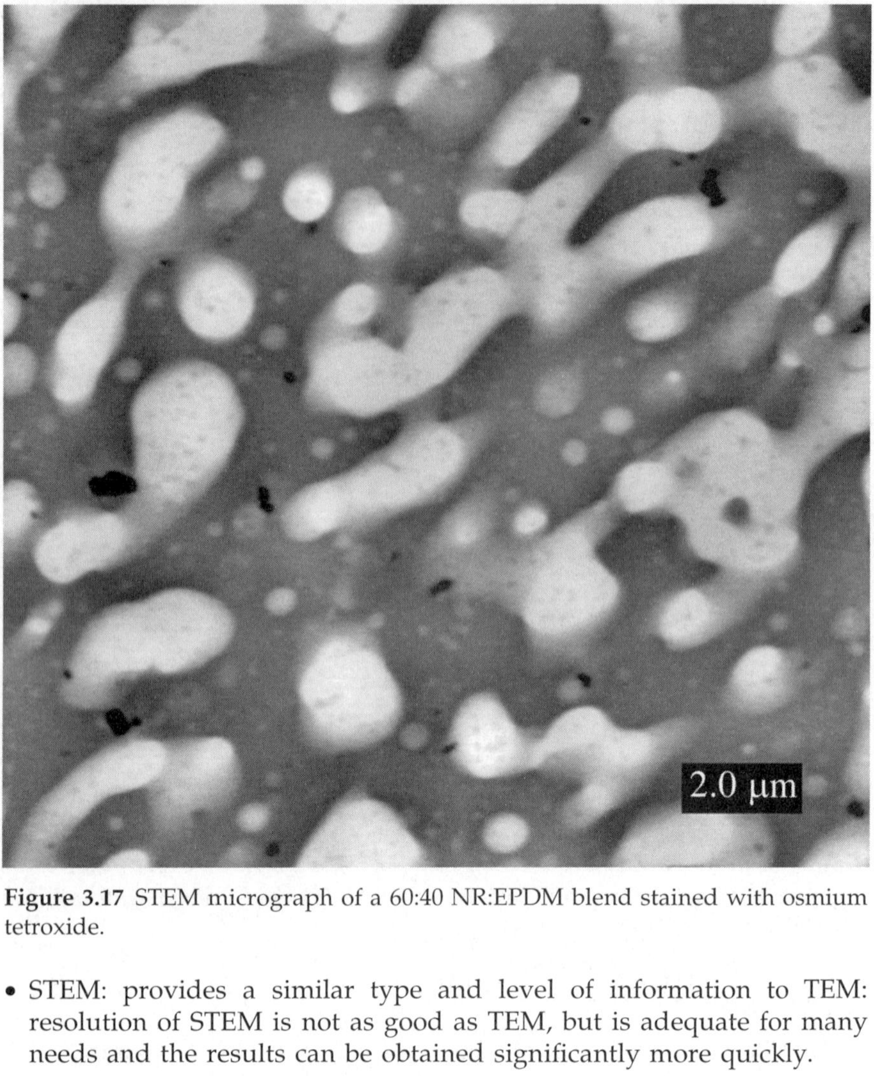


**Figure 3.17** STEM micrograph of a 60:40 NR:EPDM blend stained with osmium tetroxide.

- STEM: provides a similar type and level of information to TEM: resolution of STEM is not as good as TEM, but is adequate for many needs and the results can be obtained significantly more quickly.

## 3.9 REFERENCES

1 Sawyer, L.C. and Grubb, D.T. (1987) *Polymer microscopy*, Chapman and Hall, London.

2 Cudby, P.E.F. and Gilbey, B.A. (1995) Scanning transmission imaging of elastomer blends using an unmodified conventional scanning electron microscope. *Rubber Chemistry and Technology*, **68**(2), 342–350.

3 Hemsley D.A. (1987) *Applied polymer light microscopy*, Elsevier, London, pp. 60–65.

4 Haynes, R. (1984) *Optical microscopy of materials*, International Textbook Co., London, pp. 71–76.

5 Swayer, L.C. and Grubb, D.T. (1987) *ibid.*, p. 16.
6 Sawyer, L.C. and Grubb, D.T. (1987) *ibid.*, p. 28.
7 Hobbs, S.Y. and Watkins, V.H. (1982) The use of chemical contrast in the SEM analysis of polymer blends. *Journal of Polymer Science, Polymer Physics Edition*, **20**(4), 651–658.

# 4

# Estimation of crosslink density by solid-state NMR spectroscopy

*Aris bin Ahmad and Abu bin Amu*
*Rubber Research Institute of Malaysia, Malaysia*

## 4.1 INTRODUCTION

The technique for estimating the extent of crosslinking in the individual phases of a blend by using peak broadening associated with increasing crosslink density, as described in Chapter 2, was developed on an instrument which is designed to obtain spectra of materials in a liquid state. Thus it is necessary to swell vulcanizates for several days before endeavouring to obtain a spectrum. The work described in this chapter has sought to establish whether the technique for estimating crosslink densities in the component rubbers of a vulcanized blend can be translated to a solid-state NMR spectrometer by adopting appropriate operating procedures, and thereby facilitate the estimation of crosslink densities in vulcanized blends.

Carbon $^{13}C$ NMR spectroscopy is very useful for structure elucidation of polymeric compounds in solution. However, by using a combination of techniques like magic angle spinning, high power proton decoupling and cross polarization, it has become possible to obtain liquid-like spectra from solid polymers, e.g. vulcanized rubbers, at room temperature. There are two ways of obtaining a spectrum of polymers; namely gated high power decoupling (hpdec) and cross polarization (cpmas). Hpdec is best for examining the mobile region of polymers and cpmas is best for the less mobile region.

*Blends of Natural Rubber*. Edited by Andrew J. Tinker and Kevin P. Jones.
Published in 1998 by Chapman & Hall, London. ISBN 0 412 819406

The width of peaks is dependent upon a number of factors, but for polymers one of the more important is the mobility of the polymer chains: the higher the mobility, the sharper the peaks. Crosslinking a rubber through vulcanization decreases the mobility of the chains, and hence leads to increased peak width. Thus, there are evident differences between the spectra of lightly and highly crosslinked natural rubber vulcanizates. The width of the peak can be quantified and correlated with crosslink density. Peak line–shape, described as the width of peak at half height in units of Hertz (Hz) is one of the means for measuring crosslink density. In this work, cpmas was not used as the decrease in mobility of the polymers is still measurable by hpdec, and the effect of crosslinking, particularly for NR, is observable by hpdec.

## 4.2 EXPERIMENTAL PROCEDURES

Polymers used were gum vulcanizates of natural rubber (NR), ethylene-propylene rubber of 34% and 54% propylene content (EPM 034 and EPM 054 respectively), ethylene-propylene terpolymer rubber (EPDM), 50 mole% epoxidized natural rubber (ENR-50) and nitrile rubber (NBR) of 34% and 45% acrylonitrile content. Vulcanization was based on dicumyl peroxide for all rubbers, unless otherwise specified. The compounds were prepared on a two-roll mill. The curing characteristics of the compounds were assessed at 150 °C using a Monsanto ODR 2000 rheometer. Sheets, 2 mm thick, were compression moulded and vulcanized in a press at 150 °C to optimum cure ($t_{90}$). The elastic constant, $C_1$, was determined from equilibrium modulus measurements performed on a Greensmith's machine [1]. The values of the physically manifested crosslink concentration were then calculated via the Mooney–Rivlin relationship.

Vulcanizate samples were cut into very small pieces and packed into the rotor which was made of zirconium oxide, size 20.0 mm long (with lid) and 6.95 mm outside diameter. $^{13}$C NMR solid-state spectra were recorded on a Bruker AMX-400WB spectrometer operating at 23–25 °C and 100.16 MHz. The chemical shifts were adjusted by using tetramethylsilane (TMS) as an external secondary reference (signal at 0.00 ppm). Single pulse excitation (SPE) experiments were performed using a "train" of closely spaced 90° pulses (5.3 μsec), a delay of 5 sec, high power proton decoupling (47 Hz), and magic angle spinning.

## 4.3 NMR SPECTRA OF NR

The chemical shift positions of the signals in the solid-state $^{13}$C NMR spectrum of NR are given by Zaper and Koenig [2] as:

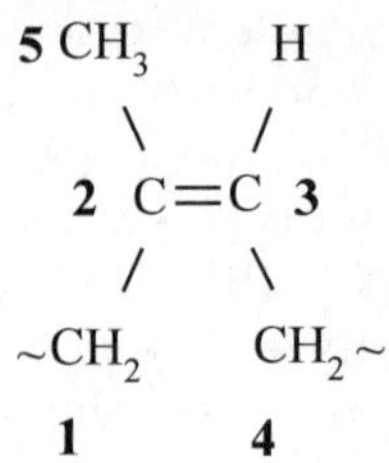


| *Carbon designation* | *Chemical shift, ppm (approx)* |
|---|---|
| C–1 | 33.7 |
| C–2 | 135.9 |
| C–3 | 126.5 |
| C–4 | 27.8 |
| C–5 | 24.7 |

It was necessary to identify the most suitable signal or signals in the spectrum for the measurement of peak broadening and to define a general procedure for its measurement. The signals due to both the olefinic and aliphatic carbons of NR were examined, peak broadening being assessed indirectly as the line width at half height of the peaks ($W_{1/2}$). Figures 4.1 and 4.2 show portions of the spectra at various chemical shift positions for the NR vulcanizates. The spectra obtained showed a visible broadening effect in the samples with increasing levels of peroxide.

The effect of dicumyl peroxide levels on the physically manifested crosslink concentration of NR are shown in Table 4.1. Figure 4.3 shows the relationship between physical crosslink density of selected vulcanizates and $W_{1/2}$ for olefinic signals, and Figure 4.4 shows the same for aliphatic signals.

It can be seen that the line width at half height increases with increasing crosslinking in NR. The results of $W_{1/2}$ for the aliphatic signals were very scattered compared to the results for the olefinic signals. Hence, the olefinic portion of the spectra appeared to exhibit a more distinct correlation between line width at half height against physical crosslink density of the NR. When spectra for the remainder of the vulcanizates were evaluated, only the olefinic signals were considered. Unidentified additional peaks were observed for NR with very high peroxide levels (20 phr dicumyl peroxide: Figure 4.1), and the spectrum was not evaluated. The completed calibration plot showing the dependence of $W_{1/2}$ for the two olefinic carbons on physical crosslink density is reproduced in Figure 4.5, where carbon atoms C–2 and C–3 have been defined earlier.

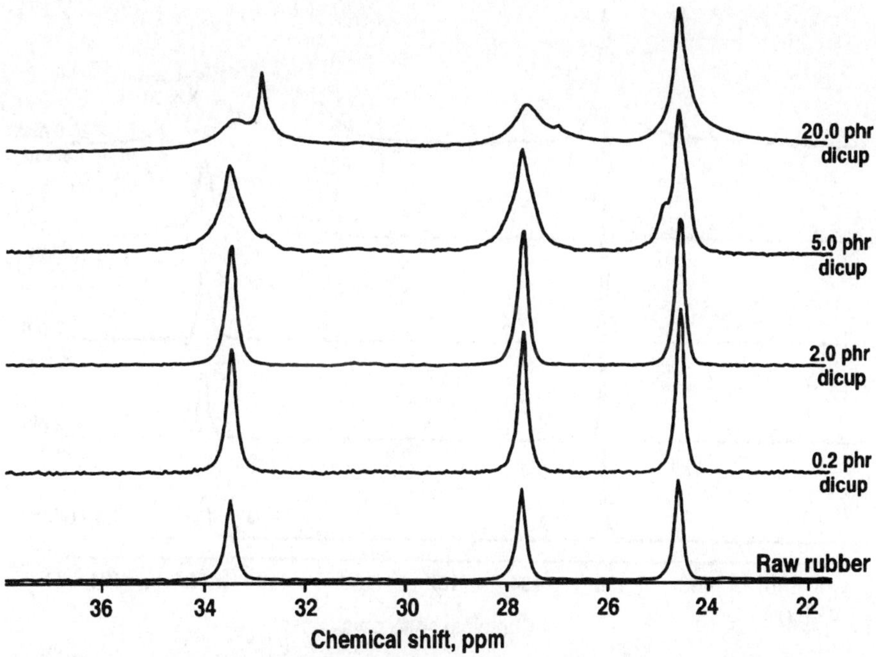


**Figure 4.1** Solid-state $^{13}C$ NMR spectra of NR gum vulcanizates cured with different levels of dicumyl peroxide; aliphatic region.

At this juncture, it was necessary to replace the probe in the NMR spectrometer. This and accompanying changes in operating practice resulted in changes in line width observed for the NR vulcanizates cured with dicumyl peroxide, as may be seen by comparing Figures 4.5 and 4.6. Line widths are smaller, but a correlation between line width and crosslink density remains – as it should. This experience is expected; any correlation of line width and crosslink density will be peculiar to a particular instrument configuration and operating practice. It serves to emphasize the importance of using fixed operating conditions when performing this type of measurement, as is well established for the swollen-state NMR technique.

To determine if similar effects can be observed for sulphur curing systems in NR, compounds were prepared and vulcanized by using a range of sulphur and TMTD levels at a fixed S:TMTD ratio. The formulation used is shown in Table 4.2. The effect of the increasing sulphur/TMTD levels on the physical crosslink density of NR are shown in Table 4.3. Solid-state NMR spectra of these vulcanizates were obtained, and the observed line widths of the olefinic signals are included in Table 4.3. Figure 4.6 includes the relationships between physical cross-

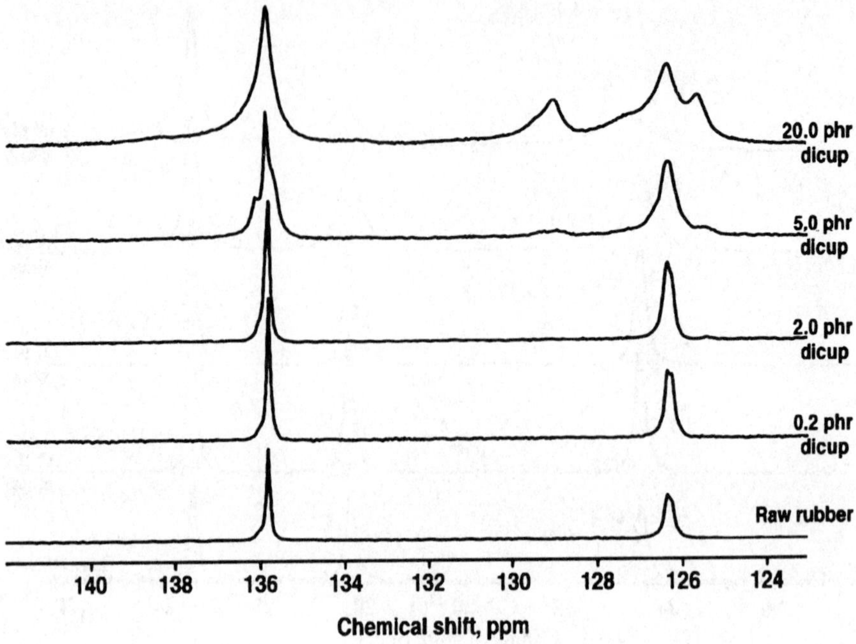


**Figure 4.2** Solid-state $^{13}C$ NMR spectra of NR gum vulcanizates cured with different levels of dicumyl peroxide; olefinic region.

**Table 4.1** Effect of dicumyl peroxide on crosslink density of NR

| *Dicumyl peroxide (phr)* | *Physical crosslink (mol/m$^3$)* |
|---|---|
| 0.25 | 5 |
| 0.5 | 13 |
| 0.75 | 16 |
| 1.0 | 18 |
| 1.5 | 26 |
| 2.0 | 33 |
| 3.0 | 47 |
| 4.0 | 62 |
| 5.0 | 70 |
| 20.0 | 267 |

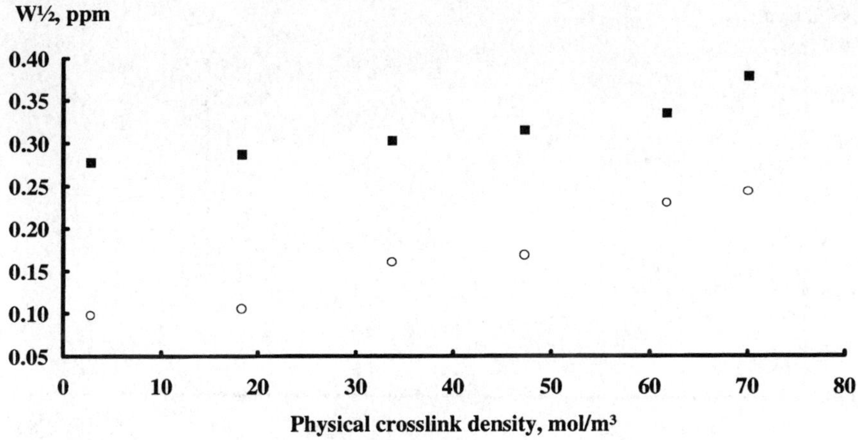


**Figure 4.3** Dependence of line width at half height ($W_{1/2}$) on physical crosslink density for NR; C–2 (○) and C–3 (■) olefinic signals.

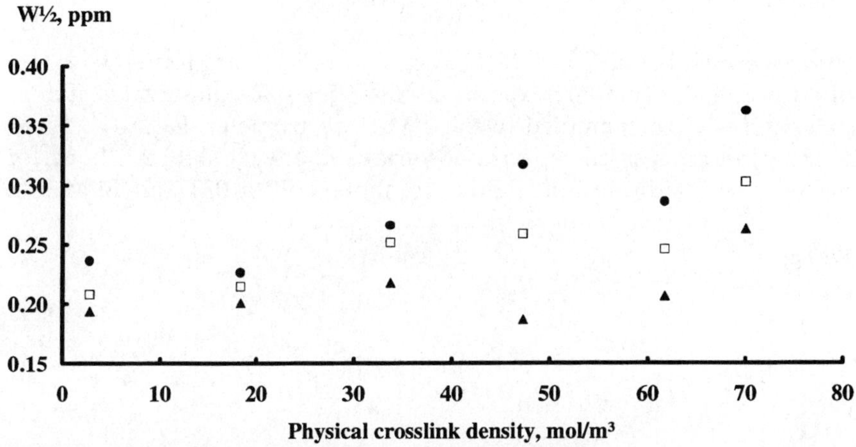


**Figure 4.4** Dependence of line width at half height ($W_{1/2}$) on physical crosslink density for NR; C–1 (●), C–4 (□) and C–5 (▲) aliphatic signals.

link density and $W_{1/2}$ for both peroxide and sulphur gum vulcanizates of NR. These results indicate that sulphur vulcanizates of NR fit quite well onto the same calibration curve as for peroxide vulcanizates.

## 4.4 NMR SPECTRA OF OTHER RUBBERS

Table 4.4 shows the physical crosslink densities of ENR-50, EPM 034, EPDM and NBR vulcanizates used in this study. For each rubber the increase in physical crosslinks with increasing peroxide levels is very

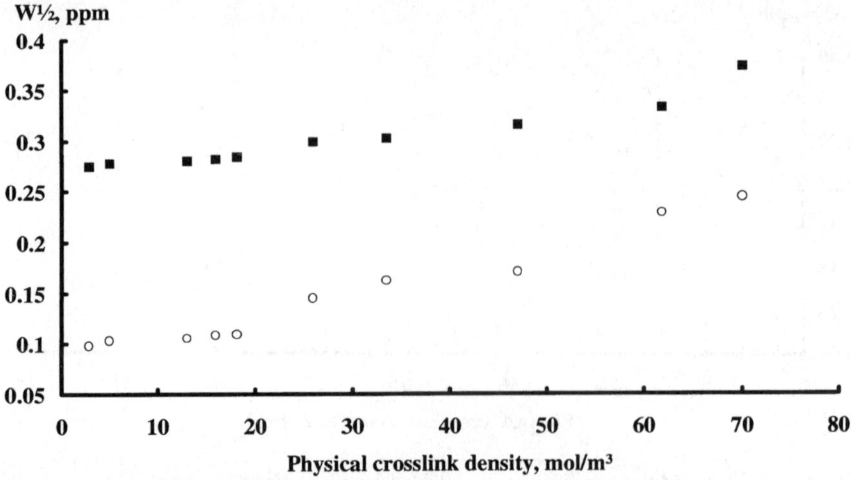


**Figure 4.5** Dependence of line width at half height ($W_{1/2}$) on physical crosslink density for NR; C–2 (○) and C–3 (■) olefinic signals.

distinctive. Solid-state $^{13}C$ NMR spectra of EPM vulcanizates were acquired under the standard conditions used for NR vulcanizates. Triplicate samples were examined in the NMR spectrometer. Figures 4.7 and 4.8 show portions of the spectra at various chemical shift positions for selected low, medium and highly crosslinked EPM 034 and EPM 054,

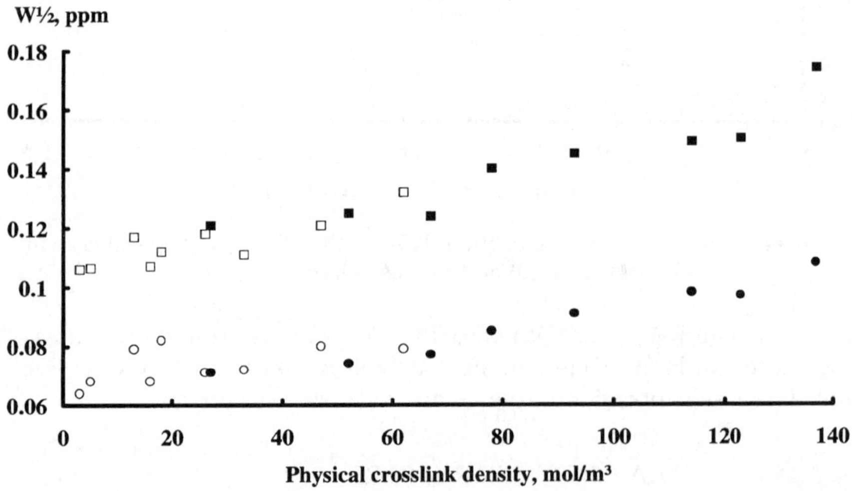


**Figure 4.6** Dependence of line width at half height ($W_{1/2}$) on physical crosslink density for NR peroxide cured [C–2 (○) and C–3 (□)] and sulphur cured [C–2 (●) and C–3 (■)] olefinic signals.

**Table 4.2** Formulation of NR compounds for sulphur vulcanization

| *Ingredient* | *phr* |
|---|---|
| SMR L | 100 |
| Stearic acid | 2 |
| Zinc oxide | 5 |
| TMQ | 1.5 |
| Sulphur | 0.5–4.0 |
| TMTD | 0.2–1.6 |

**Table 4.3** Effect of increasing sulphur/TMTD levels on physical crosslink density

| *Physical crosslink ($mol/m^3$)* | *Line width at half height (ppm)* | |
|---|---|---|
| | *C–2* | *C–3* |
| 27 | 0.071 | 0.121 |
| 52 | 0.074 | 0.126 |
| 67 | 0.077 | 0.125 |
| 78 | 0.085 | 0.142 |
| 92 | 0.091 | 0.145 |
| 114 | 0.099 | 0.149 |
| 123 | 0.098 | 0.150 |
| 136 | 0.108 | 0.174 |

**Table 4.4** Effect of peroxide level on physical crosslink density for various rubbers

| *Sample number* | *Dicup (phr)* | *Physical crosslink ($mol/m^3$)* | | | | |
|---|---|---|---|---|---|---|
| | | *ENR-50* | *EPM 034* | *EPDM* | *Krynac 34.50*[a] | *Krynac 45.55*[b] |
| 1 | 0.25 | 5 | 1 | 18 | 33 | 27 |
| 2 | 0.5 | 11 | 9 | 38 | 70 | 62 |
| 3 | 0.75 | 12 | 17 | 60 | 96 | 85 |
| 4 | 1.0 | 25 | 21 | 68 | 128 | 129 |
| 5 | 1.5 | 32 | 29 | 92 | 206 | 191 |
| 6 | 2.0 | 42 | 34 | 112 | 390 | 255 |
| 7 | 3.0 | 53 | 50 | 128 | 396 | 279 |
| 8 | 4.0 | – | 61 | 154 | – | – |
| 9 | 5.0 | – | 71 | 180 | – | – |

[a] NBR, 34 wt% acrylonitrile, Bayer
[b] NBR, 45 wt% acrylonitrile, Bayer

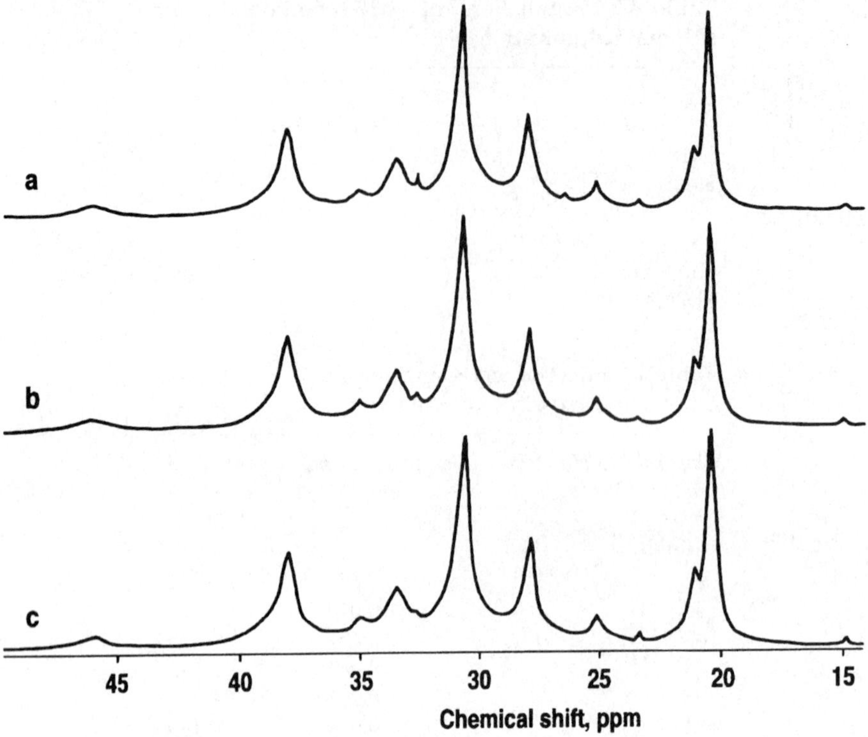


**Figure 4.7** Solid state $^{13}C$ NMR spectra of EPM (Copolymer 034) peroxide vulcanizates with high (a), medium (b), and low (c) crosslink densities.

respectively. The spectra show a very slight decrease in line width with increasing crosslink density. The actual values of the line width at half height of the spectra for both copolymers (with sample numbers 1, 5 and 9 for low, medium and high crosslinks respectively) are shown in Table 4.5. Of the many peaks in the spectrum, those at 37 ppm (designated as peak for C–1) and 30 ppm (peak for C–2) were suitable for calculating $W_{1/2}$.

Figure 4.9 shows the relationship between physical crosslink density and $W_{1/2}$ for the designated aliphatic carbon peaks for EPM 034. In general, these results show an apparent decrease in $W_{1/2}$ with increasing crosslinking level, although the decrease in values is rather insignificant in comparison with the changes seen for NR vulcanizates. These changes observed for EPM did not appear large enough to be utilized for calibration purposes.

The spectra obtained for ENR-50 (Figure 4.10) show very little peak broadening with increasing crosslink levels when measured at room temperature. This was thought to be due to the difference in the tem-

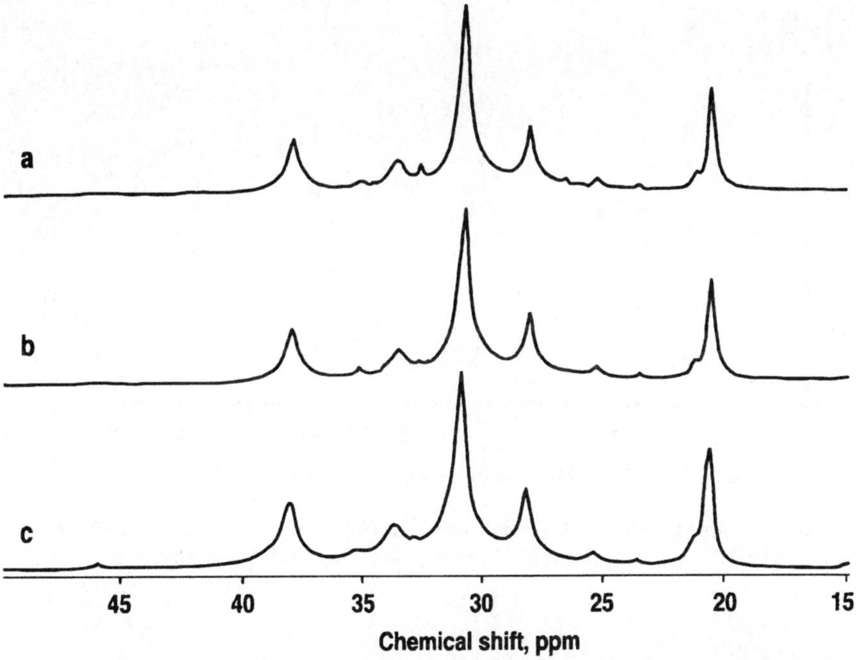


**Figure 4.8** Solid state $^{13}C$ NMR spectra of EPM (Copolymer 054) peroxide vulcanizates with high (a), medium (b), and low (c) crosslink densities.

perature at which the spectra were recorded relative to the glass transition temperature of the rubber. Measurements were then made at 35 °C, 40 °C, 45 °C, 50 °C and 100 °C. The spectra reproduced in Figure 4.11 show examples of the effect of temperature on the spectrum of a particular ENR-50 vulcanizate with a specific crosslink level. There is

**Table 4.5** Line width at half height in $^{13}C$ NMR spectra of EPM peroxide vulcanizates

| *Sample number* | *EPM 034* | | *EPM 054* | |
|---|---|---|---|---|
| | *C–1*[a] | *C–2*[b] | *C–1*[a] | *C–2*[b] |
| 1 | 0.710 | 1.026 | 0.091 | 0.778 |
| 5 | 0.628 | 0.813 | 0.605 | 0.743 |
| 9 | 0.672 | 0.882 | 0.591 | 0.691 |

[a] 37 ppm
[b] 30 ppm

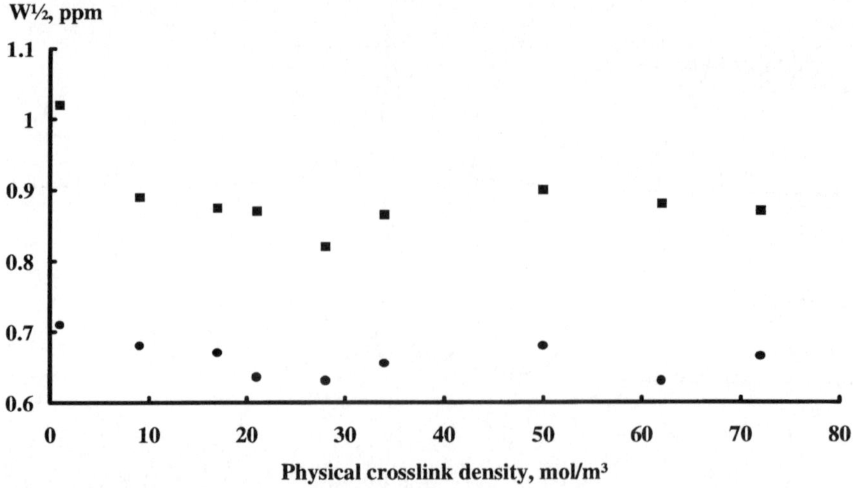


**Figure 4.9** Dependence of C–1 (●) and C–2 (■) line width at half height ($W_{1/2}$) on physical crosslink density for EPM (Copolymer 034) peroxide vulcanizates.

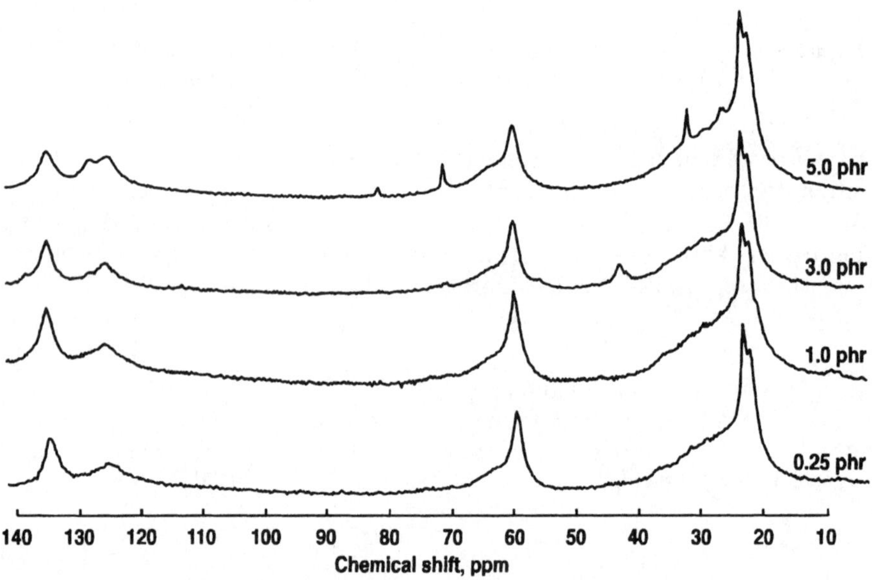


**Figure 4.10** Solid-state $^{13}C$ NMR spectra of ENR-50 gum vulcanizates cured with different levels of dicumyl peroxide; room temperature NMR.

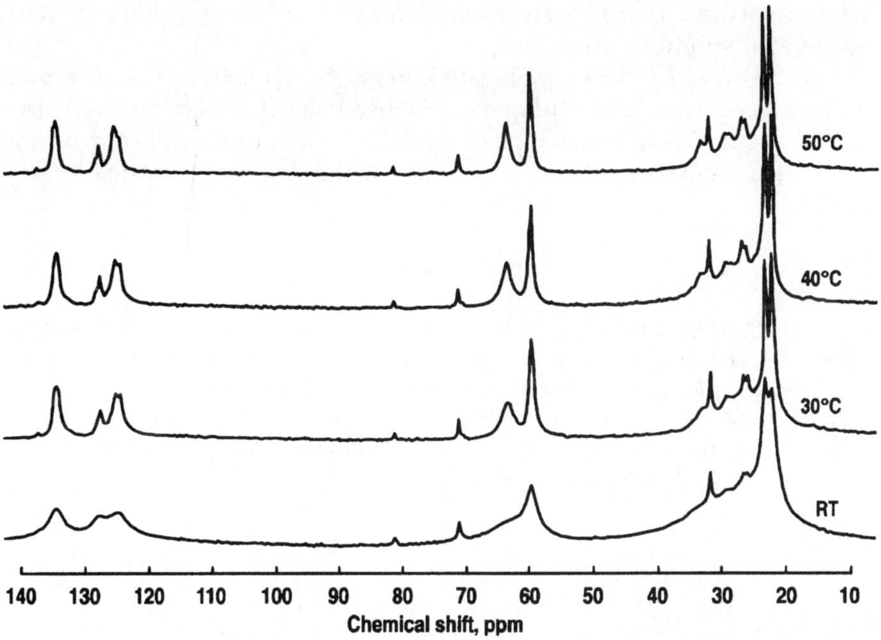


**Figure 4.11** Solid-state $^{13}C$ NMR spectra of ENR-50 gum vulcanizates cured with different levels of dicumyl peroxide; variable temperature NMR.

an abrupt change in the signal width at temperatures just above ambient, but little change thereafter. Similar behaviour was observed for vulcanizates cured with a higher level of peroxide, and there was no dependence of peak width on crosslink density at the higher temperatures. Consideration was given to changing the spectrometer conditions such as spinning rate and an increase in the number of scans, but no improvement was observed.

Spectra of NBR gum vulcanizates were also measured, and similar observations as for ENR-50 vulcanizates were seen. There appears to be a slight peak shift to higher field with increased crosslink densities, which is not easy to explain. The changes are again too small for utilization for calibration.

## 4.5 CONCLUSION

It has been shown that, for gum NR vulcanizates, there is a correlation between physical crosslink density and line width at half height, which increases with increasing crosslink density. The correlation is nearly linear and is meaningful for quite a broad range of crosslink densities.

The same correlation is seen for vulcanizates cured with peroxide and accelerated sulphur cure systems.

For the other rubbers investigated, no such correlation was observed despite using a similar technique to that used for NR. There may be some limitations to this approach, or possibly the correct techniques may not have been devised for these elastomers within the very restricted time available.

## 4.6 REFERENCES

1 Campbell, D.S., Chapman, A.V., Goodchild, I.R. and Fulton, W.S. (1992). Experimental determination of the Mooney–Rivlin constant for natural rubber vulcanizates. *Journal of Natural Rubber Research*. **7**(3) 168.

2 Zaper, A.M. and Koenig, J.L. (1986). Solid state carbon-13 NMR studies of vulcanized elastomers. II. Sulfur vulcanization of natural rubber. *Rubber Chemistry and Technology*, **60**(2), 252–277.

# 5

# NR/NBR blends – basic problems and solutions

*Michael V. Lewan*
*Tun Abdul Razak Research Centre, MRPRA, UK*

## 5.1 INTRODUCTION

Nitrile rubber (NBR) is a copolymer of acrylonitrile and butadiene. It is the acrylonitrile component which makes NBR an oil-resisting rubber; the higher its proportion then the greater the oil resistance. Two nitrile rubbers were studied: one containing 41% acrylonitrile, Breon N41 (Zeon Chemicals), and a second containing 34% acrylonitrile, Krynac 34.50 (Bayer). These will be referred to as NBR41 and NBR34 respectively.

The blending together of natural rubber (NR) and NBR is intended to produce a vulcanizate with the best properties from each component, i.e. NBR's high resistance to swelling by oils and NR's good strength properties. In practice the result is often disappointing in terms of properties and service life. Causes include a maldistribution of crosslinks which results in an over-crosslinked phase and a poorly crosslinked phase. For blends of two rubbers differing in polarity, such as NR and NBR, maldistribution of crosslinks can arise through preferential solubility of the curatives and vulcanization intermediates. Previous work had shown that the distribution of crosslinks can be controlled by the selection of the cure system [1].

Estimation of crosslink densities in vulcanized blends is crucial to identifying appropriate cure systems which minimize either partitioning of curatives or the subsequent effect of partition on the eventual crosslink distribution achieved during vulcanization. Crosslink densities in the individual phase of NR/NBR blends have been estimated routinely in this

*Blends of Natural Rubber*. Edited by Andrew J. Tinker and Kevin P. Jones.
Published in 1998 by Chapman & Hall, London. ISBN 0 412 819406

work by swollen-state $^1$H NMR spectroscopy using a conventional liquids spectrometer (300 MHz Bruker AC300), as described in Chapter 2. The technique is based on correlations of spectral peak width with physical crosslink density, that is, the physically effective crosslink density – $n_{phys}$.

The extent of the crosslinking across the interface between NR and NBR is another problem. The difference in polarity of the rubbers causes high interfacial tension, which is detrimental. It will severely limit mixing at the interface, and hence the opportunity for crosslinking between the rubbers. It also causes poor phase morphology, which is characterized by large phase sizes. Network visualization microscopy was used to study the interface as described in Chapter 2. Interfacial adhesion was also investigated by subjecting the blends to differential swelling: this technique examines the consequences of stresses caused by swelling the major phase of a blend in a solvent which does not swell the minor phase [2]. The work to be described was also influenced by the ability of being able to estimate crosslink densities in blends – the results were only interpretable with this information [3].

The investigation initially examined blends of NR with the higher acrylonitrile content NBR, i.e. NBR41. Work focused particularly on curatives approved by the US Food and Drugs Administration (FDA) because there was an ultimate interest in food contact applications.

## 5.2 NR/NBR41

### 5.2.1 Cure systems

Cure systems for 50:50 NR:NBR41 blends were investigated in gum vulcanizates to establish those which give an even distribution of crosslinks in both phases of a blend. The main accelerators investigated were the commonly used sulphenamides – CBS (N-cyclohexylbenzothiazole-2-sulphenamide), MBS (N-oxydiethylenebenzothiazole-2-sulphenamide) and TBBS (N-*t*-butylbenzothiazole-2-sulphenamide). Although these all belong to the same family of accelerators, the differing amine substituents might be expected to have some influence on crosslink distributions. Semi-efficient vulcanization cure systems were selected for the initial work. A 1:1 sulphur:accelerator ratio was chosen at a level of 1.3 phr for the compounds based on CBS; the same molar ratio was used for the other sulphenamide accelerators. All compounds were prepared by crossblending previously mixed NR and NBR masterbatches containing 5 phr zinc oxide, 2 phr stearic acid and 1.5 phr TMQ antioxidant on a two-roll mill and adding the curatives.

The sulphenamide cure systems are satisfactory for press cures, but higher cure rates are required for injection moulding. Therefore, the effect of small amounts of secondary accelerators on both cure rate and crosslink distribution were investigated. The secondary accelerators chosen were TMTM (tetramethylthiuram monosulphide), DPG (diphenyl guanidine), TBTD (tetrabutylthiuram disulphide), MBTS (2,2′-benzothiazole disulphide) and ODIP (N,N′-dioctadecyl N,N′-diisopropylthiuram disulphide). The level of secondary accelerator selected was 0.1 phr for TMTM, and the same molar level for the other secondary accelerators.

The blends were cured to rheometer $t_{max}$ (maximum rise in rheometer torque) at 150 °C. The crosslink densities ($n_{phys}$) for both rubber phases were determined by swollen-state NMR spectroscopy. The CBS accelerator on its own gives a maldistribution of crosslinks in favour of NR (Figure 5.1).

Addition of TMTM to the cure system over-compensated, with the crosslinks now favouring the NBR phase. ODIP increased the maldistribution in favour of NR, whilst DPG, MBTS and TBTD had little, if any, effect on the distribution of the crosslink density. The contrary action of the two closely related thiuram secondary accelerators, TMTM and ODIP, may be rationalized in terms of their structure and the effect on partition in this blend system. Sulphur will be preferentially soluble in (partition in favour of) the more polar elastomer [4]. The implication of the distribution observed in the absence of secondary accelerator is that CBS partitions in favour of NR in this particular NR/NBR blend. TMTM is a polar species and is clearly partitioning in favour of the NBR. ODIP has larger alkyl substituents than TMTM and is consequently less polar; to the extent that it appears to partition in favour of NR.

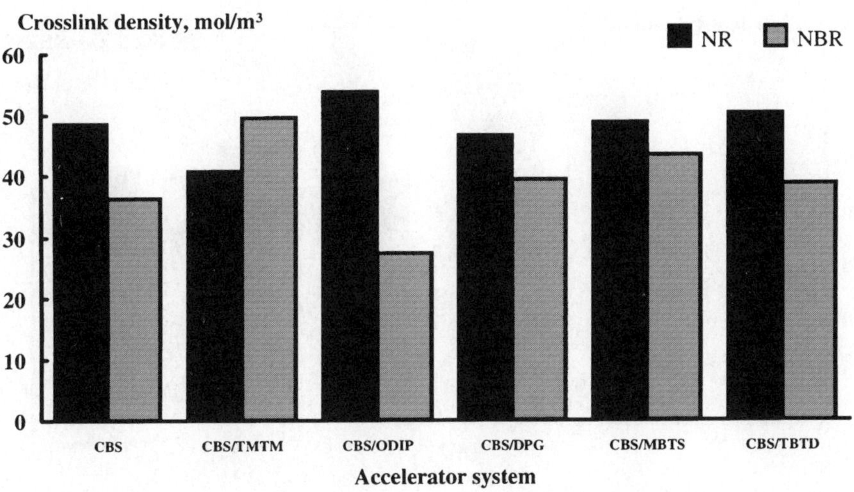


**Figure 5.1** Crosslink density distributions in 50:50 NR:NBR41 blends cured at 150 °C.

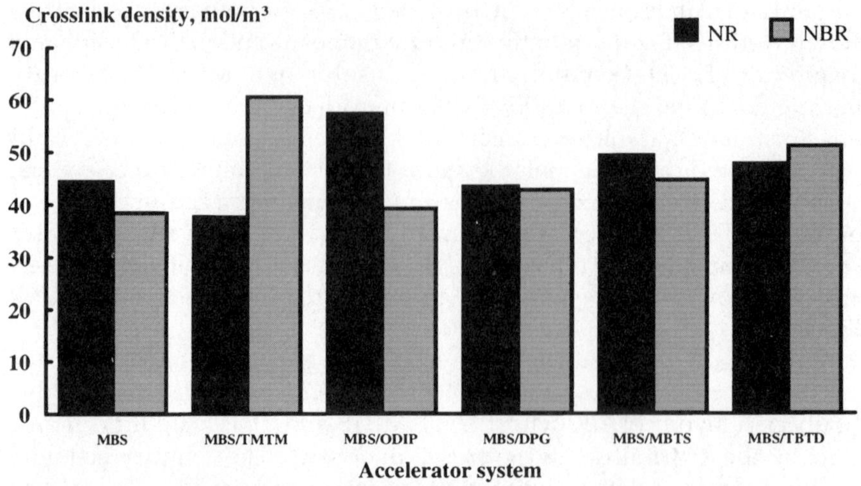


**Figure 5.2** Crosslink density distributions in 50:50 NR:NBR41 blends cured at 150 °C.

MBS on its own gives a fairly even distribution of crosslinks, but at a significantly lower overall crosslink density (Figure 5.2).

The addition of the secondary accelerators had a similar effect to that observed in the CBS accelerated systems. TBBS on its own produced an even distribution of crosslinks between the two rubber phases. Neither DPG nor TBTD changed the distribution (Figure 5.3). Adjusting the level

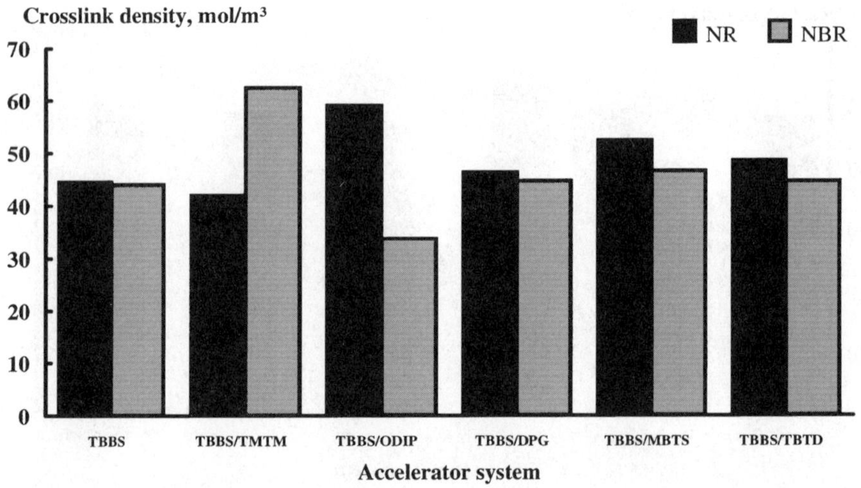


**Figure 5.3** Crosslink density distributions in 50:50 NR:NBR41 blends cured at 150 °C.

of some of the secondary accelerators produced further blends which had a more equal distribution of crosslinks between the rubber phases.

These results show that similar accelerators can produce quite different crosslink distributions. Even rather similar sulphenamide accelerators cannot be exchanged with impunity. TBBS appears to be a promising candidate for press curing and the cure rate can be boosted by TBTD and MBTS without adversely affecting crosslink distribution.

Blends were also cured to rheometer $t_{90}$ (90% rise to rheometer torque) and this gave cure times of between 10 and 20 minutes. The crosslink densities and tensile strengths of these blends were similar to those cured at $t_{max}$.

The tensile strength of a vulcanized blend is affected by factors other than crosslink distribution: phase size and overall crosslink density may be expected to be influential, for instance. It is not surprising that comparison of tensile strength with the ratio of physical crosslink densities in the two rubber phases shows a very scattered picture, but there is a distinct overall pattern (Figure 5.4). The highest tensile strengths are evident around a NR/NBR crosslink density ratio of about 1.0–1.2, i.e. close to even crosslink density distribution between the two phases. Ratios either side of this central band tend to cause a decline in tensile strength, especially when the maldistribution of crosslinks favours the NBR phase. Slight maldistribution favouring the NR phase is acceptable. Thus the inherently stronger rubber, NR, must have a sufficient crosslink density for the blend to have good strength properties. The highest tensile strengths are given by blends vulcanized with TBBS as the sole

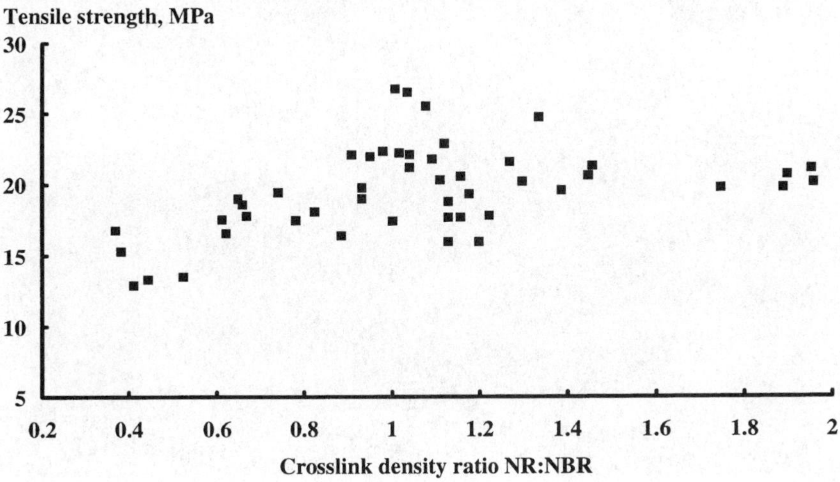


**Figure 5.4** Dependence of tensile strength on crosslink density ratio for 50:50 NR:NBR41 gum blends cured at 150 °C.

accelerator. Values in excess of 20 MPa are noteworthy for a difficult blend, particularly an unfilled blend.

### 5.2.2 Phase morphology

The blends were examined initially by optical (phase contrast) microscopy. Sections about 1 μm thick were taken by cryomicrotomy and micrographs obtained as described in Chapter 3 (Figure 5.5). Large phase sizes were observed, as expected given the large differences in polarities, and hence solubility parameters, between the two rubbers [5]. The micrographs also suggest that the phase texture is co-continuous, although the NBR (darker phase) appears to have a more discrete character. The

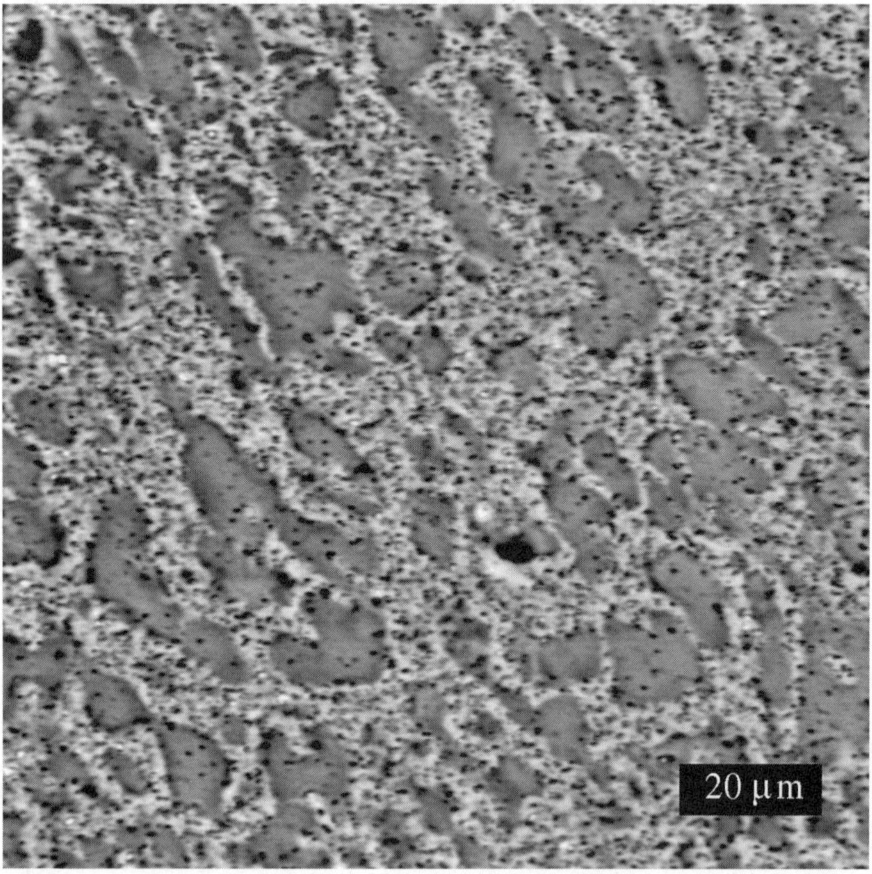


**Figure 5.5** Phase contrast micrograph of a 50:50 NR:NBR41 gum blend vulcanizate.

particles of zinc oxide (black spots) appear to be preferentially, although not exclusively, located in the NR phase.

Modification of the sample mount for the scanning electron microscope (SEM) has produced a new procedure, namely STEM (scanning transmission electron microscopy) based on SEM (Chapter 3). Sections for analysis by this technique are stained with osmium tetroxide. STEM is sufficiently sensitive to variations in osmium level that the NR and NBR are clearly distinguishable despite both phases being reactive toward the stain. A STEM micrograph (Figure 5.6) clearly shows that the NBR phase (lighter regions) contains smaller regions of NR, and this might be a factor in the better than expected strength properties of these blends.

There are also instances of failure (white areas) at the interface during sectioning at low temperatures. The preferential location of zinc oxide (black dots) in the NR phase is confirmed.

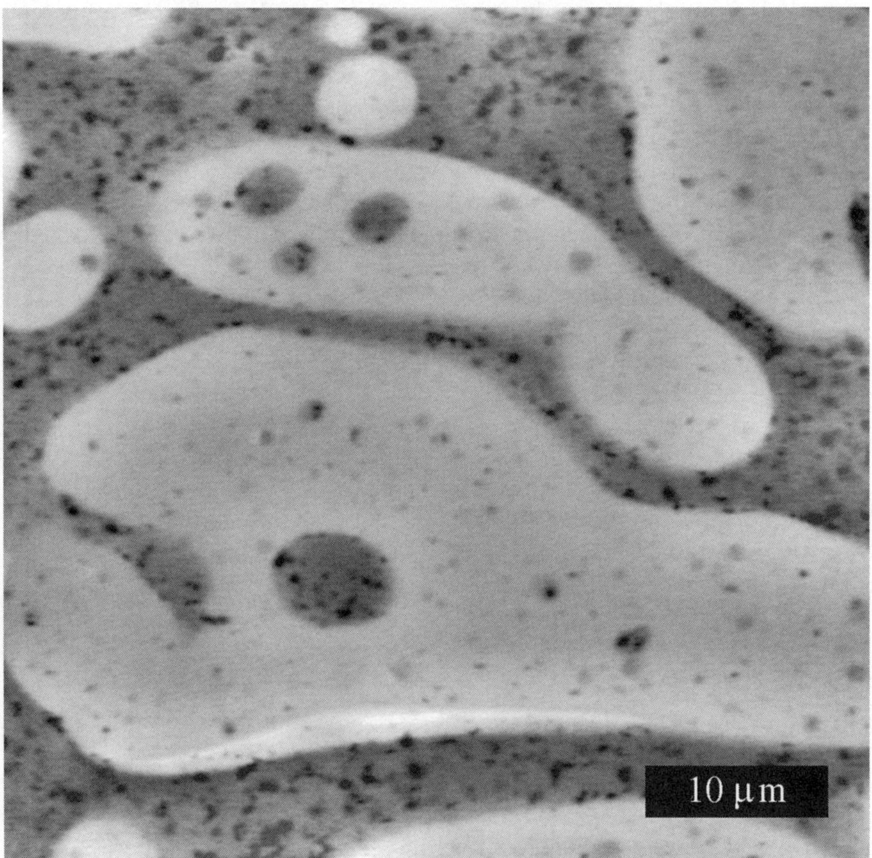


**Figure 5.6** STEM micrograph of a 45:55 NR:NBR41 gum blend vulcanizate.

### 5.2.3 Interfacial strength

The crosslinking between the two phases can only occur within the mixed interphase. Thus, there is some possibility that there will be insufficient linking between the phases, especially given the limited mixing to be expected from two polymers of such differing polarities and solubility parameters. This weakness could lead to premature failure in service.

*(a) Network visualization*

The nature of the interface between the two rubbers was also investigated by another new microscopy technique, 'network visualization', which is described in Chapter 2. The vulcanizate is swollen to equilibrium in styrene and the styrene is then polymerized. Ultra-thin sections are taken from the resulting composite and stained with osmium tetroxide prior to viewing by transmission electron microscopy (TEM). A mesh structure comprising strands of rubber in a polystyrene matrix is revealed and the mean size of the mesh cells is correlated with the molecular weight between crosslinks, $M_c$. Relatively large regions of polystyrene accumulate at weaknesses within the vulcanizate – around zinc oxide particles for instance, or in the case of blends at weak interfaces. A micrograph of a section prepared from a blend which has low tensile strength and a maldistribution of crosslinks in favour of the NBR component shows extensive separation at the interface (Figure 5.7). The band of polystyrene approximates up to 250 nm in thickness with only occasional strands of network material linking the two rubbers. This suggests less crosslinking between the two rubbers than within either phase.

A micrograph of a section prepared from a 50:50 NR:NBR41 gum vulcanizate which has high tensile strength, and hence is believed to have adequate interfacial adhesion, is shown in Figure 5.8. There is some separation at the interface between the rubbers, as there is a concentration of polystyrene in a band about 30 nm thick bridged by network material indicative of crosslinks between the two rubbers.

*(b) Differential swelling*

Interfacial adhesion was also investigated by another technique: blends were subjected to differential swelling in which the major phase of the blend is swollen in a solvent that does not swell the minor component [2, 3]. For instance, *iso*-octane swells NR but not NBR, whilst acetone swells NBR but not NR. Two swelling terms are required: $V_{ro}$, which represents the swelling of the rubber to be expected without the non-swelling component, and $V_r$, representing the actual swelling of the

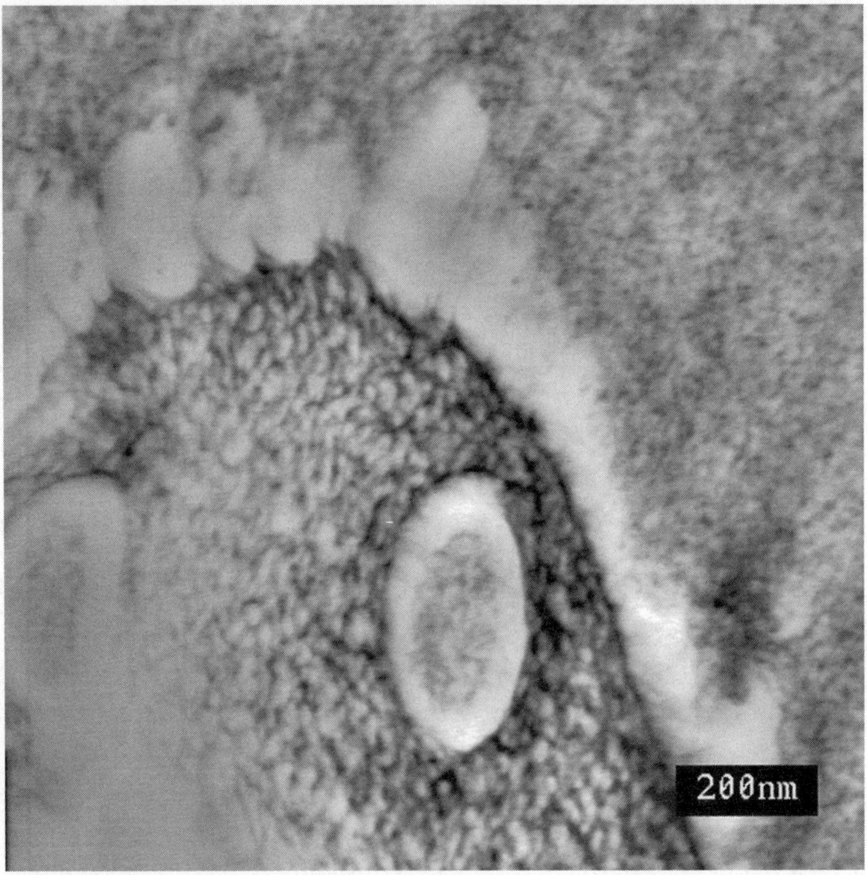


**Figure 5.7** TEM micrograph of a 50:50 NR:NBR41 gum blend vulcanizate prepared for 'network visualization'.

blend. Both terms are given by the volume fraction of rubber in the swollen gel. $V_{ro}$ is estimated from correlations of peak width in swollen-state NMR spectra and $V_r$ in the solvent concerned. When $V_{ro}/V_r < 1$ there is restriction of swelling, which arises from adequate bonding between the two phases resisting the stresses caused by the differential swelling and thereby constraining swelling. When $V_{ro}/V_r > 1$, interfacial failure has occurred; the stresses incurred in differential swelling caused interfacial failure and solvent filled the resulting voids between the rubber phases. Thus, greater swelling is observed.

NR/NBR blends were produced with ratios of 75:25, 65:35, 35:65 and 25:75. The NR-rich blends were swollen in *iso*-octane whilst the NBR-rich blends were swollen in acetone. The differential swelling behaviour is

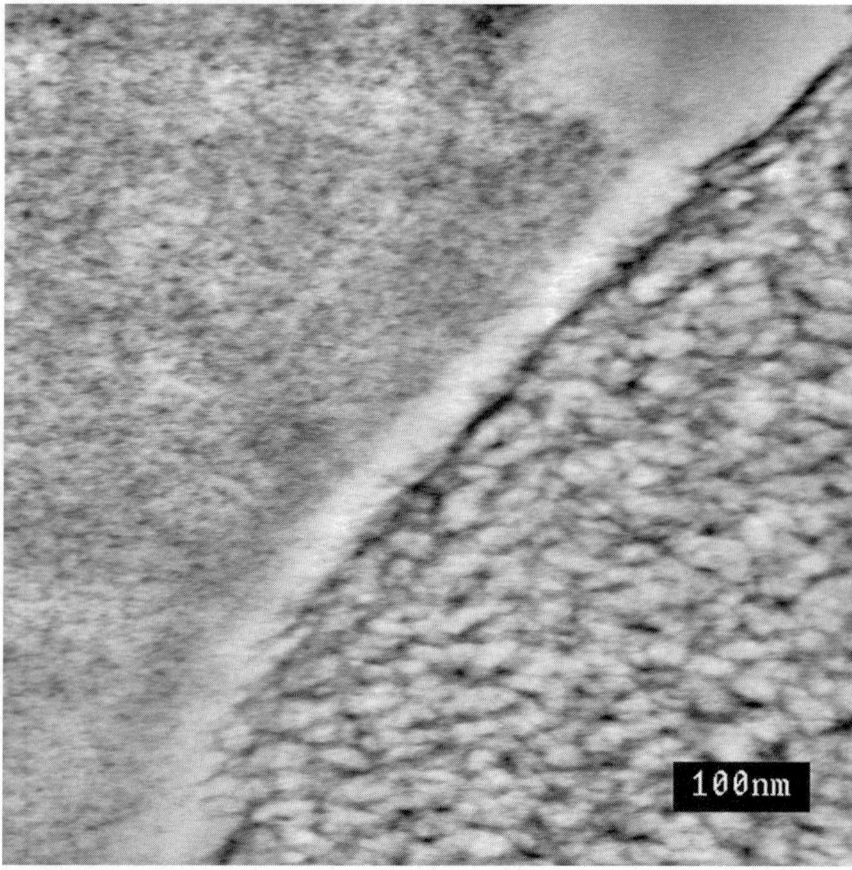


**Figure 5.8** TEM micrograph of a 50:50 NR:NBR41 gum blend vulcanizate prepared for 'network visualization'.

presented within the context of tensile strength of the blend vulcanizates in Figure 5.9. Values on the left-hand side of the plot represent good interfacial adhesion ($V_{ro}/V_r < 1$) and those on the right ($V_{ro}/V_r > 1$) represent poor interfacial adhesion. The blends with good interfacial adhesion also have good tensile strengths, but as the interfacial adhesion lessens so does the tensile strength. Three bands are observed. The band of good interfacial adhesion and high tensile strength is represented by the NR-rich blends: this particularly high strength stems from the strain-crystallizing NR matrix. The middle band consists of the 35:65 NR:NBR blends, and the third band comprises the 25:75 NR:NBR blends. In considering these observations, it must be borne in mind that there is a progressive decrease in overall crosslink density with increasing NBR

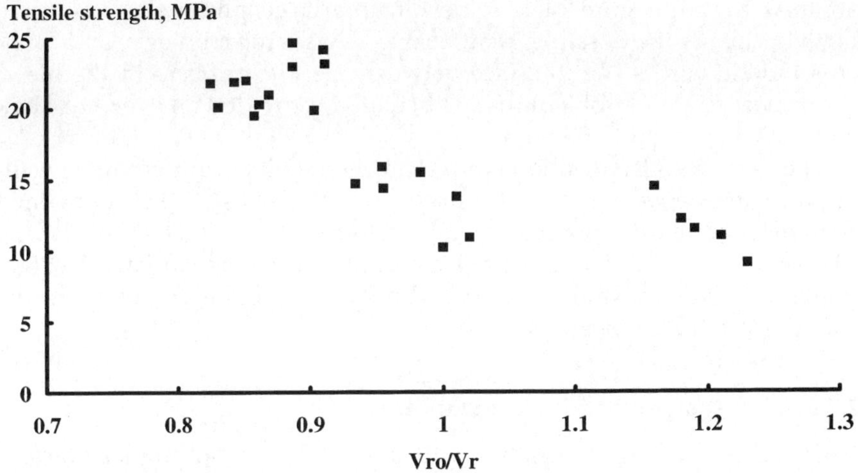


**Figure 5.9** Dependence of tensile strength of NR/NBR41 gum blend vulcanizates on swelling restriction ratio.

content of the blends – as illustrated in Figure 5.10. This will inevitably reduce the level of crosslinking between the two rubbers across the phase boundary.

Increasing the sulphur level in the NR/NBR blends which had poor interfacial adhesion led to increases in the crosslink density of each phase and improvement in interfacial strength. The substantial tensile strengths

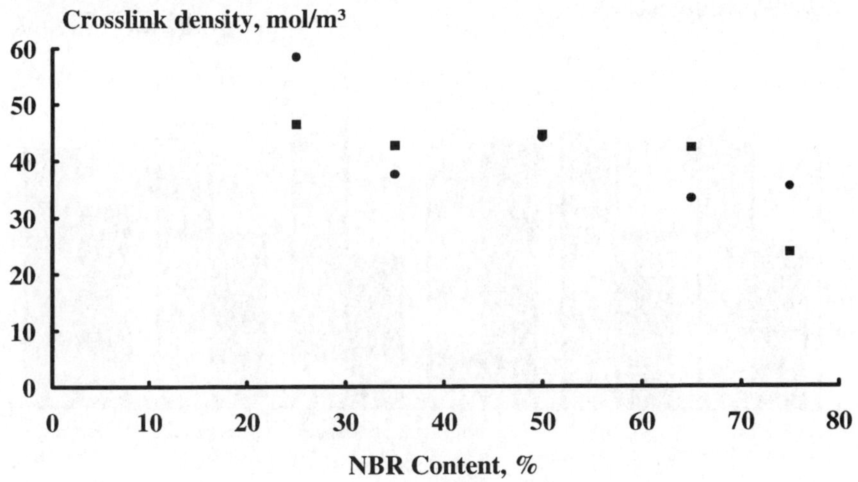


**Figure 5.10** Dependence of crosslink density in NR and NBR41 phases of gum blend vulcanizates (S/TBBS) in NBR41 content.

attained by these gum blends, despite the large phase sizes, is again noteworthy. It is possible that coarse phase morphology and poor crosslinking across the interface between the two rubbers in the blend will cause less of a problem than anticipated, provided an even crosslink distribution is maintained.

The network visualization study and the results of differential swelling measurements suggest that adequate interfacial strength is obtained provided there is no marked maldistribution of crosslinks between the two rubbers and that the overall crosslink density is sufficiently high. Both conditions maximize the opportunity for crosslinking between the rubbers at the interface.

### 5.2.4 High temperature vulcanization

Vulcanization at 150 °C has been considered so far, but higher vulcanization temperatures are required in processes like injection moulding. Batches of the blends previously considered were vulcanized at 180 °C to maximum cure as indicated by a Monsanto MDR2000E rheometer, and crosslink densities were estimated by swollen-state $^1$H NMR spectroscopy. Apart from the blends vulcanized with ODIP as secondary accelerator, the NBR phase has a greater crosslink density than the NR phase under these conditions – as illustrated in Figure 5.11 for cure systems based on CBS.

Compared with the blends cured to $t_{max}$ at 150 °C, those cured at 180 °C have lower overall crosslink densities. At 180 °C the NR phase is

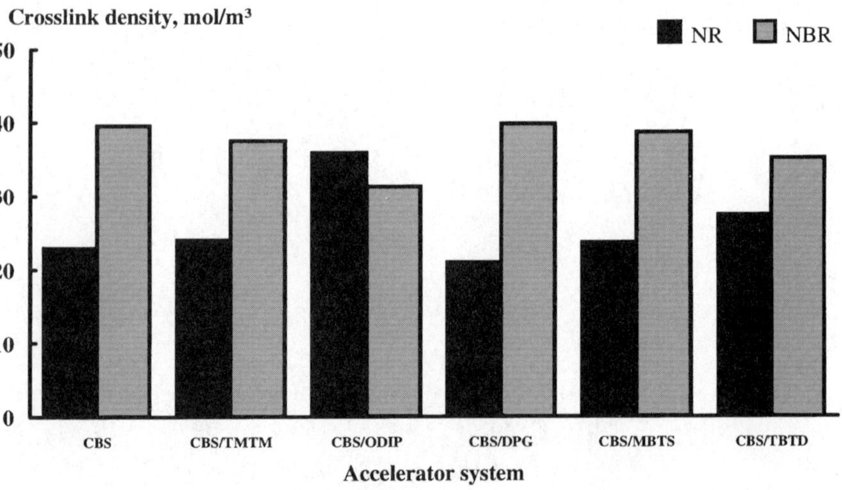


**Figure 5.11** Crosslink density distributions in 50:50 NR:NBR41 blends cured at 180 °C.

significantly lower in crosslink density. Crosslink densities in NBR vary from being very similar to those achieved at 150 °C to being much lower, especially when the crosslink density at 150 °C was very high. A probable explanation for this behaviour is the efficiency of the cure system operating within the NBR phase. This will be governed by the relative partitioning of sulphur and accelerator(s).

Due to the general bias of crosslinks in favour of the NBR phase, there is no clear picture of the dependence of tensile strength on the ratio of the physical crosslink densities in the two phases.

## 5.3 NR/NBR 34 BLENDS

Economic considerations favour the use of a lower acrylonitrile content NBR. Thus, Krynac 34.50 was considered in 50:50 blends with NR vulcanized with the three common sulphenamide accelerators and cured at both 150 °C and 180 °C. The crosslink densities of the rubber phases in each blend all showed heavy biasing towards the NBR phase at 150 °C (Figure 5.12), which is accentuated at 180 °C.

The crosslink density of the NR phase in all blends is lower by at least 10 mol/$m^3$ at 180 °C compared to 150 °C, whilst the crosslink density of the NBR phase varies from showing virtually no change to large reductions. This overall fall in crosslink density yield is reflected in the much poorer tensile strengths. Only two of the blends have reasonable strengths, namely those based on CBS/TBTD and CBS/TETD cure systems: both have very high tensile strengths (ca. 25 MPa) at 150 °C. The

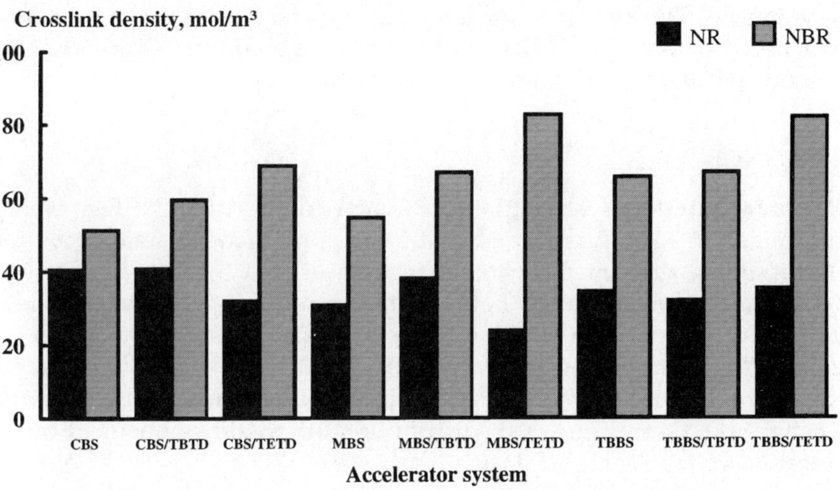


**Figure 5.12** Crosslink density distributions in 50:50 NR:NBR34 blends cured at 150 °C.

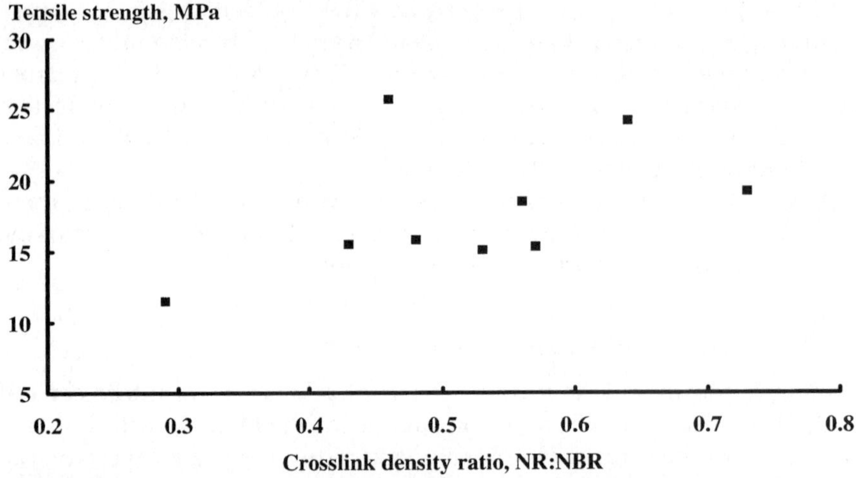


**Figure 5.13** Dependence of tensile strength on NR:NBR crosslink density ratio for 50:50 NR:NBR34 gum blends vulcanized at 150 °C.

relationship between tensile strength and the ratio of physical crosslink densities in the two phases is not very clear in this case, but as all of the crosslink distributions are skewed towards the NBR phase any trend may be difficult to detect (Figure 5.13).

The difficulty in identifying a cure system capable of giving an even distribution of crosslinks in this blend system may be attributed to the lower solubility parameter of an NBR with 34% rather than 41% acrylonitrile. The probable explanation is that none of the accelerators considered within the study would partition in favour of NR to offset the effect of partition of sulphur in favour of the NBR.

## 5.4 CONCLUSION

Adequate interfacial strength is obtained in 50:50 NR:NBR41 blends vulcanized at 150 °C, provided that there is no marked maldistribution of crosslinks between the two rubber phases and the overall crosslink density is sufficiently high. Both of these conditions serve to maximize the opportunity for crosslinking between the rubbers at the interface. A semi-EV S/TBBS cure system is suitable, and the cure rate may be enhanced by the use of TBTD or MBTS as secondary accelerators without significantly altering the even distribution of crosslinks obtained in the vulcanizate.

At injection moulding temperatures of 180 °C, lower overall crosslink densities are produced, most notably in the NR phase, than in the case of blends cured at 150 °C. This could be overcome by increasing curative levels. In most cases the highest tensile strengths are given by blends which have the highest crosslink density in the NR phase. The tensile strengths of the blends when the NR phase is adequately crosslinked are very high, even for the unfilled vulcanizates considered herein.

Lower overall levels of crosslinking are observed in NR/NBR41 blends of lower NR content, and the interfacial strength is low. To overcome this, the level of sulphur may be increased to increase the overall crosslink density, and thus promote adequate crosslinking across the interface.

50:50 blends of NR with the lower acrylonitrile rubber, Krynac 34.50, vulcanized with accelerators approved by the FDA did not produce an even distribution of crosslinking between the phases, but tensile strengths were adequate for vulcanizates cured with either S/CBS/TBTD or S/CBS/TETD semi-EV cure systems.

## 5.5 REFERENCES


1 Lewan, M.V. (1996) *Crosslink density distribution in NR/NBR blends*. PhD Thesis, Loughborough University.

2 Zapp, R.L. (1973) Chlorobutyl covulcanization chemistry and interfacial elastomer bonding. *Rubber Chemistry and Technology*, **46**(1), 251–274.

3 Tinker, A.J. (1990) Crosslink distribution and interfacial adhesion in vulcanized blends of NR and NBR. *Rubber Chemistry and Technology*, **63**(4), 503–515.

4 Gardiner, J.B. (1968) Curative diffusion between dissimilar elastomers and its influence on adhesion. *Rubber Chemistry and Technology*, **41**(5), 1312–1328.

# 6

# Improving the morphology and properties of NR/NBR blends with NR/PMMA graft copolymers

*K.G. Karnika de Silva* and Michael V. Lewan*[†]
**Rubber Research Institute of Sri Lanka, Sri Lanka;*
*[†]Tun Abdul Razak Research Centre, UK*

## 6.1 INTRODUCTION

Blends of highly incompatible elastomers may sometimes be improved by the addition of small amounts of another polymer. Setua and White [1] applied this technique to improve the homogeneity of binary and ternary blends of CR, NBR and EPM. When a small amount of chlorinated polyethylene is added to NBR/EPM or CR/EPDM blends, they mix more rapidly. The chlorinated polyethylene acts as a compatibilizing agent which appears to form a skin on the EPM particles that helps the larger NBR chunks adhere to them. This increased adhesion and polarity at the EPDM surface increases the compatibility. The presence of block or graft copolymers can also alleviate blending of incompatible elastomers as they can alter interfacial properties [2–6]. Ideally the block or graft component should contain a segment which is chemically identical to one of those in the respective phases, but the desired effect may still be achieved if one polymer of the graft is miscible with, or adhered to, one of the phases.

*Blends of Natural Rubber*. Edited by Andrew J. Tinker and Kevin P. Jones.
Published in 1998 by Chapman & Hall, London. ISBN 0 412 819406

It is possible to obtain considerable improvement in the phase morphology and tensile properties of NR/NBR gum blend vulcanizates by incorporating typically 5–20% of methyl methacrylate grafted natural rubber (e.g. Heveaplus MG30), or chloroprene rubber (CR) (Chapter 7). MG30 is a mixture of graft copolymer with ungrafted NR and poly(methyl methacrylate), PMMA, present as a homopolymer. It was considered that there would be sufficient interaction between the PMMA graft chains and the acrylonitrile repeat units of the NBR for the graft copolymer to locate at the NR/NBR interface and thus reduce interfacial tension and hence phase size.

A number of mixing schemes were investigated to evaluate how they affect the overall homogeneity and phase morphology of the different components of the elastomer blends. Work was also performed to identify curing systems which would give even crosslink distributions within the blends with compatibilizers, since successful curing systems would not necessarily be the same as those found for uncompatibilized NR/NBR blends. The type and percentage of the compatibilizer, viscosities of the elastomers [7], blending technique [7–9], the temperature of mixing and the type of accelerator appear to have a major influence on phase morphology, and hence the homogeneity and the physical properties of the blends.

## 6.2 METHODS OF BLENDING

Three grades of NBR are considered here – Breon N41C80 and N41C45 (Zeon Chemicals) containing 41% acrylonitrile and with Mooney viscosities of 80 and 45 respectively, and Perbunan N28C45 (Bayer) containing 28% acrylonitrile and with a Mooney viscosity of 45. These will be referred to as NBR41C80, NBR41C45 and NBR28C45 respectively.

Three mixing procedures were used:

1. Separate masterbatches of NR, MG30 and NBR, containing 5 phr zinc oxide, 2 phr stearic acid and 1.5 phr TMQ, were prepared in a BR Banbury internal mixer and the masterbatches were cross-blended and finalized with curatives on a two-roll mill.
2. NR and MG30 were cross-blended and masterbatched with zinc oxide, stearic acid and TMQ in a BR Banbury internal mixer and cross-blended with the NBR masterbatch and finalized with curatives on a two-roll mill.
3. NR, MG30 and NBR were cross-blended in a Brabender PL2000 Plasticorder fitted with a 350S mixing head and Banbury rotors at a rotor speed of 100 rpm for 3 minutes. Zinc oxide, stearic acid, TMQ and curatives were added in a second stage mixing for 2.5 minutes at a circulating oil temperature of 40 °C and a rotor speed of 80 rpm. The oil temperature in the first mixing stage was 50 °C, 100 °C or 110 °C.

Test sheets (2 mm thick) were vulcanized in a press at 150 °C for the time to maximum torque rise ($t_{max}$) as estimated from rheographs obtained using a Monsanto MDR 2000E rheometer operating at the same temperature. In addition to measurements of physical properties, crosslink densities were estimated by swollen-state $^1$H NMR spectroscopy, using a General Electric QE300 spectrometer, as described in Chapter 2. The phase morphology of NR/NBR41 blends has been investigated previously by optical microscopy, but there are difficulties, due to low contrast between the phases, and so STEM, a modified form of SEM, was used instead (Chapter 3).

## 6.3 PHASE MORPHOLOGY

For a 65:35 NR:NBR41C80 blend (Figure 6.1a), the observations confirm the large phase sizes obtained in the absence of MG30 rubber and illustrate the weak interfacial adhesion (white areas) wherein a region of the two phases appears to have been pulled apart from each other during sectioning. This may be a significant factor in the low tensile strength observed for this blend. It has also been noted that there appears to be a reasonable amount of NR as a darker microphase structure within the lighter NBR phase. For both a 50:50 MG:NBR41 blend (Figure 6.1b) and a 10:50 blend containing 40 parts MG30 (Figure 6.1c) two co-continuous phases are seen, and both of these exhibit fine structures. The lighter microphase in the NR phase is probably largely unstained PMMA in MG30; this will exist as microphases as in MG30 itself [10]. The darker phase in the NBR phase is probably an NR microphase, as seen in Figure 6.1a.

Micrographs of blends of NR/NBR28C45, with and without MG30 rubber, are depicted in Figure 6.2. As discussed earlier, it has been found that the morphology of immiscible rubber blends is dependent on the mixing procedure, the rheological properties of the blend components and their degree of compatibility. There is a greater incidence of co-continuity in the phases when the NBR has a low acrylonitrile content with a similar viscosity to SMR L. Large variation in phase size (in cross-section) of the continuous NR phase within each sample is clearly seen, even with very low levels of MG30. As might be expected, the mean phase sizes appear to decrease with increasing MG30 content, such that the largest phase size is obtained with the blend which has no added compatibilizer (Figure 6.2a). The NBR phase appears to be discrete in each case except as shown in Figure 6.2d, where the morphology appears to be co-continuous: this takes no account of the morphology outside the plane of the section in which continuity of the NBR phase may be preserved. In the last three samples the NR phase contains a microstructure which is presumably PMMA.

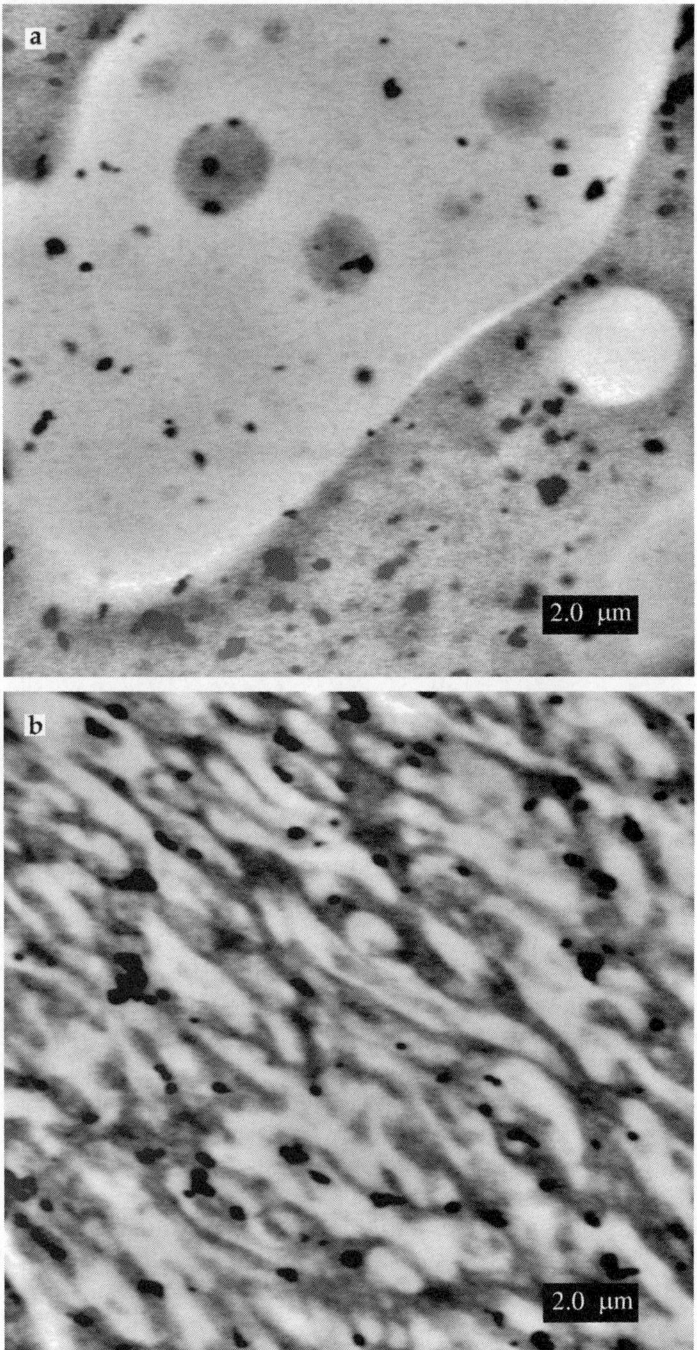


**Figure 6.1a, b**

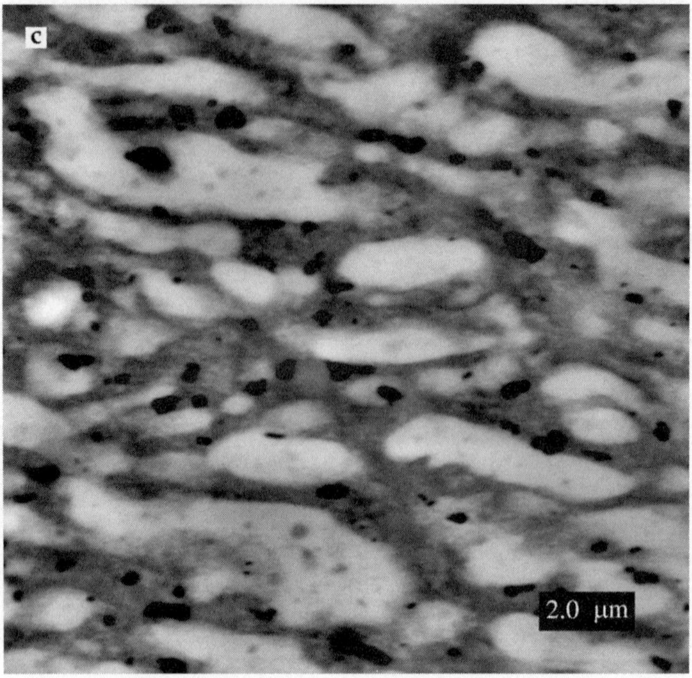


**Figure 6.1** STEM micrographs of gum vulcanizates containing; (a) 65:35 NR/NBR41 – with no compatibilizer, (b) 50:50 MG30/NBR41 – with no NR, (c) 10:40:50 NR/MG30/NBR41.

The probable reason for phase sizes in the blends containing MG30 being larger than seen in Figure 6.1b and 6.1c is the lower shear prevailing during mixing in a Plasticorder compared to mixing on a two-roll mill. The blends shown in Figure 6.1 were mixed according to Method 1, whereas those shown in Figure 6.2 were mixed according to Method 3. Cross-blending on a mill was shown to give superior properties over Mixing Procedure 3 in a Plasticorder. However, the introduction of compatibilizers produced the most significant improvements.

## 6.4 PHYSICAL PROPERTIES

The benefit of having a compatibilizer, such as MG30 rubber, in the system is evident when properties are compared with those of NR/NBR vulcanizates without compatibilizer. Examples of blends vulcanized with 1.5 phr sulphur and 0.6 phr TMTD or 1.5 phr sulphur and 1.93 phr ODIP cure systems are presented in Tables 6.1 and 6.2 respectively. Higher tensile strengths are observed in the presence of MG30 rubber

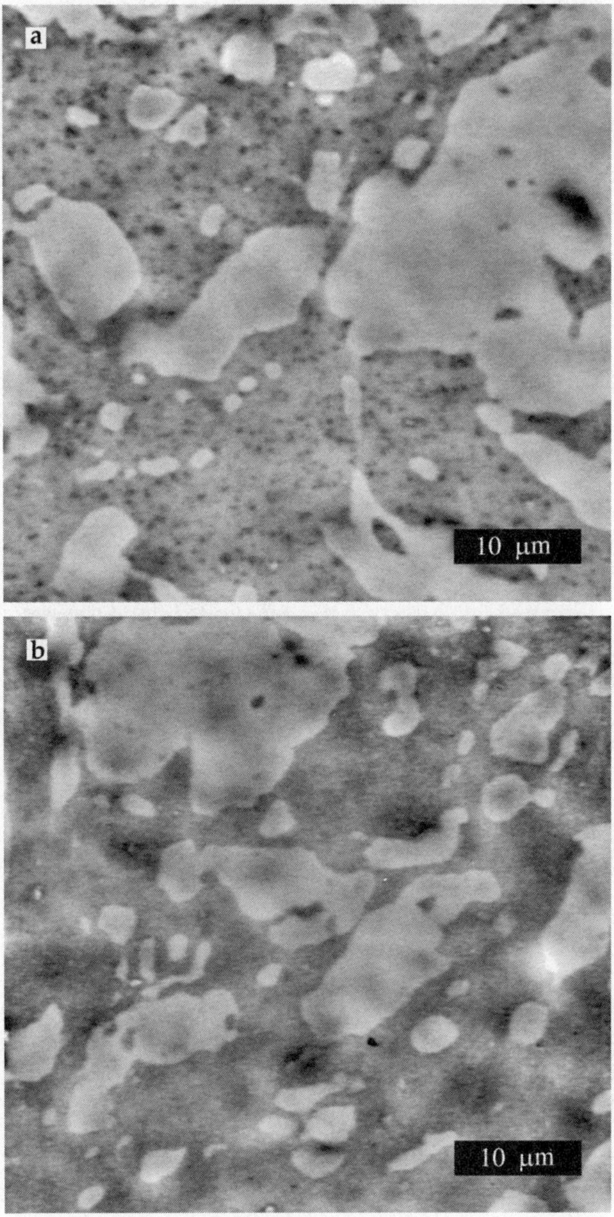


**Figure 6.2a, b**

than for comparable binary blends. Whilst part of the improvement may be attributed to a reinforcing effect of the substantial amounts of MG30, a reduction in phase size will also play a role. The higher strengths for

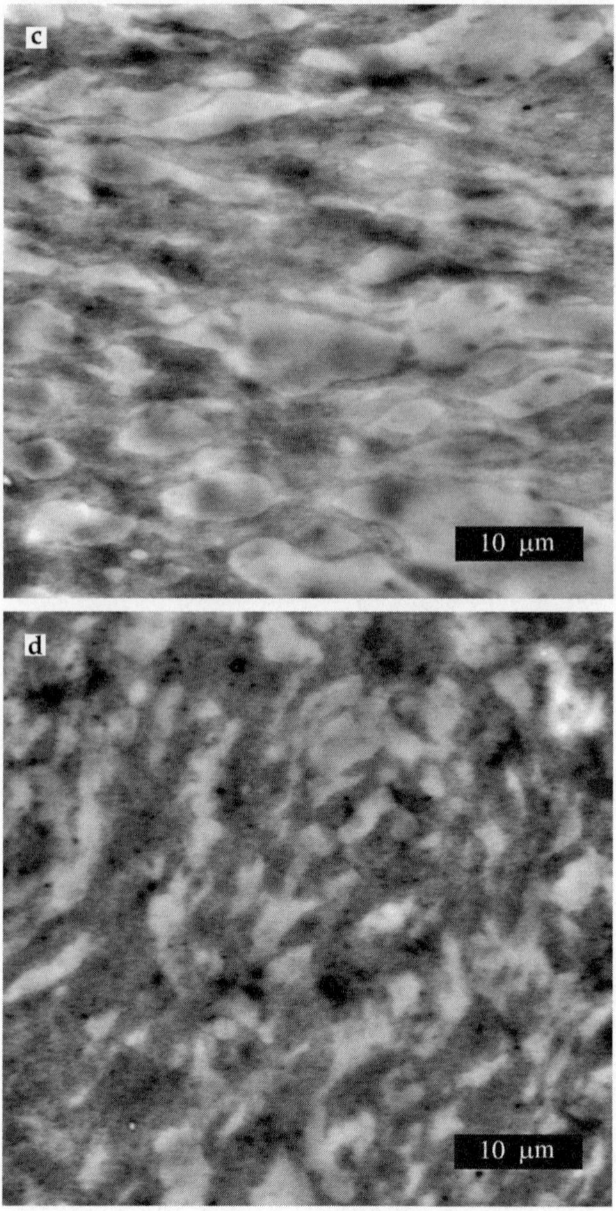


**Figure 6.2** STEM micrographs of NR/MG30/NBR28C45 gum vulcanizates (S/TBBS); (a) 50:50 NR:NBR, (b) 40:15:45 NR:MG30:NBR, (c) 25:25:50 NR:MG30:NBR, and (d) 50:50 MG30:NBR.

**Table 6.1** Physical properties of NR/NBR41C80 blends cured with S/TMTD

| *NR:MG30:NBR* | *M100 (MPa)* | *Tensile strength (MPa)* | *Elongation at break (%)* | *Trouser tear strength (N/mm)* | *Oil swell ASTM #1 (%)[a]* |
|---|---|---|---|---|---|
| 25:0:75 | 0.89 | 5.6 | 485 | 2.72 | 29 |
| 4.5:20.5:75 | 1.74 | 11.5 | 475 | 4.32 | 19 |
| 50:0:50 | 0.68 | 4.3 | 600 | 1.80 | 201 |
| 9:41:50 | 2.07 | 20.5 | 465 | 2.84 | 61 |
| 75:0:25 | 0.77 | 14.3 | 770 | 1.83 | 214 |
| 13.5:61.5:25 | 2.20 | 21.0 | 470 | 4.46 | 109 |

[a] Volume swelling after 70 h at 100 °C

vulcanizates cured with S/ODIP may be attributed to the differing crosslink distributions. TMTD is known to produce crosslink distributions which are strongly biased in favour of the NBR phase, whilst the use of ODIP, which has longer chain aliphatic substituents, favours crosslinking in the NR phase in this blend system (Chapter 5).

The common sulphenamide accelerator TBBS gives an even distribution of crosslinks in blends of NR with NBR containing 41% acrylonitrile and vulcanized at 150 °C (Chapter 5). The effects of NR:NBR polymer ratio and MG30 content on the physical properties of blends with N41C80 are demonstrated in Table 6.3. Tensile strength increases with increasing MG30 content. The decline in tensile strength with increasing NBR content noted for comparable binary NR/NBR blends in Chapter 5 is not evident when MG30 is present. Tear strength also does not show

**Table 6.2** Physical properties of NR/NBR41C80 blends cured with S/ODIP system

| *NR:MG30:NBR* | *M100 (MPa)* | *Tensile strength (MPa)* | *Elongation at break (%)* | *Trouser tear strength (N/mm)* | *Oil swell ASTM #1 (%)[a]* |
|---|---|---|---|---|---|
| 25:0:75 | 0.86 | 4.6 | 1000 | 7.88 | 28 |
| 4.5:20.5:75 | 1.50 | 14.8 | 780 | 7.31 | 12 |
| 50:0:50 | 0.78 | 12.7 | 960 | 7.68 | 89 |
| 9:41:50 | 2.17 | 13.7 | 565 | 5.50 | 37 |
| 75:0:25 | 0.70 | 15.8 | 870 | 5.01 | 205 |
| 13.5:61.5:25 | 2.37 | 18.6 | 570 | 4.04 | 78 |

[a] Volume swelling after 70 h at 100 °C

**Table 6.3** Physical properties of NR/MG30/NBR41C80 blends cured with S/TBBS (1.3/1.17 phr)

| *NR:MG30:NBR* | *M100 (MPa)* | *M300 (MPa)* | *Tensile strength (MPa)* | *Elongation at break (%)* | *Trouser tear strength (N/mm)* | *Oil swell ASTM #1 (%)*[a] |
|---|---|---|---|---|---|---|
| 60:15:25 | 0.94 | 2.14 | 24.6 | 715 | 2.75 | 114 |
| 50:15:35 | 1.00 | 2.12 | 24.9 | 705 | 3.35 | 80 |
| 40:15:45 | 1.04 | 2.16 | 24.0 | 720 | 4.40 | 66 |
| 45:05:50 | 0.83 | 1.43 | 19.9 | 805 | 11.3 | – |
| 40:10:50 | 0.90 | 1.68 | 21.5 | 780 | 10.9 | – |
| 35:15:50 | 1.11 | 2.27 | 23.6 | 710 | 3.86 | 58 |
| 30:20:50 | 1.20 | 2.65 | 24.6 | 672 | 4.07 | 49 |
| 30:15:55 | 1.16 | 2.28 | 22.4 | 690 | 4.14 | 50 |
| 20:15:65 | 1.25 | 2.31 | 23.4 | 750 | 5.45 | 38 |

[a] Volume swelling after 70 h at 100 °C

the expected decrease with increasing NBR content of the blend. There is evidence of an increase in resistance to swelling by oils with increasing MG30 content.

The very high tensile strengths which may be obtained in unfilled NR/MG30/NBR blends is demonstrated by the entries in Table 6.4. The tensile strengths of blends containing N41C45 are higher than those of comparable blends containing the higher viscosity N41C80 (Table 6.3). Although NR has a high initial viscosity, comparable to, or higher than, that of N41C80, it breaks down rapidly during mixing, producing a mismatch in viscosities which is not conductive to the development of a fine phase morphology. Smaller phase sizes have been seen in blends of

**Table 6.4** NR/MG30/NBR blends with high tensile strength[a]

| *NR:MG30:NBR* | *NBR grade* | *M100 (MPa)* | *M300 (MPa)* | *Tensile strength (MPa)* | *Elongation at break (%)* | *Trouser tear strength (N/mm)* |
|---|---|---|---|---|---|---|
| 55:05:40 | NBR41C45 | 0.89 | 1.63 | 27.1 | 760 | 4.94 |
| 50:10:40 | NBR41C45 | 0.93 | 1.93 | 27.7 | 715 | 4.01 |
| 45:05:50 | NBR41C45 | 0.90 | 1.59 | 24.2 | 740 | 4.15 |
| 40:10:50 | NBR41C45 | 0.94 | 1.75 | 25.0 | 745 | 4.30 |
| 60:15:25 | NBR28C45 | 0.87 | 2.00 | 26.6 | 700 | – |
| 30:30:40 | NBR28C45 | 1.31 | 3.99 | 27.3 | 610 | – |

[a] Mixed by method 3 and cured with S/TBBS (1.3/1.17 phr)

NR with N41C45, and this finer phase morphology may be responsible for the higher tensile strengths.

Given the higher cost of MG30 relative to NR and to a lesser extent NBR, it is noteworthy that most of the entries contain only modest levels of MG30. These results show that high physical properties are obtainable even with 5% or 10% of the compatibilizer incorporated in NR/NBR blends.

## 6.5 CROSSLINKING DENSITY DISTRIBUTION IN THE BLENDS

It has been reported previously [11] that 50:50 blends of NR and NBR with 41% acrylonitrile content have NR/NBR crosslink distributions of 11/82 and 43/10, when cured at 150 °C with 1.5 phr sulphur and 0.6 phr TMTD or 1.93 phr ODIP respectively. Similar or more extreme maldistributions have been observed in blends containing MG30; 9:41:50 NR:MG30:NBR have NR/NBR crosslink distributions of 12/75 and 64/5 respectively. The reduction in phase size due to the presence of the compatibilizer has not had a profound effect on the crosslink distributions given by these two cure systems.

On the other hand, the S/TBBS cure system which gives the desired even distribution of crosslinks in blends of NR with NBR containing 41% acrylonitrile no longer does so when MG30 is used as the compatibilizing agent (Table 6.5). Blends with part substitution of either NR or NBR with MG30 consistently show a significant bias of crosslinks in favour of the NR phase. This may be due to the reduction in phase size in the presence

**Table 6.5** Crosslink densities in NR/MG30/NBR blends cured with 1.3 phr S/1.17 phr TBBS

| *NR:MG30:NBR* | *NBR grade* | *Crosslink density, mol/m³* | |
|---|---|---|---|
| | | *NR* | *NBR* |
| 45:05:50 | NBR41C80 | 46.1 | 26.2 |
| 40:10:50 | NBR41C80 | 52.8 | 22.0 |
| 55:05:40 | NBR41C45 | 50.2 | 29.3 |
| 50:10:40 | NBR41C45 | 47.8 | 32.7 |
| 45:05:50 | NBR41C45 | 46.2 | 32.7 |
| 40:10:50 | NBR41C45 | 50.4 | 31.8 |
| 50:00:50 | NBR28C45 | 44.9 | 48.7 |
| 50:15:35 | NBR28C45 | 58.6 | 55.2 |
| 30:20:50 | NBR28C45 | 58.1 | 41.9 |
| 00:50:50 | NBR28C45 | 66.0 | 41.4 |

of MG30. The crosslink distribution in blends with such large phase sizes as are encountered in blends with NBR containing 41% acrylonitrile is largely determined by the distribution of curatives prior to vulcanization. Migration of curatives or vulcanization intermediates by diffusion during vulcanization can only influence the relatively small proportion of the blend within a few microns of the interface. When the phase size is reduced, migration can play a more significant role in determining the crosslink distribution.

The crosslink distribution observed for blends containing NBR with 28% acrylonitrile content, including the 50:50 binary blend, are not expected from the marked bias towards the NBR phase observed for comparable blends containing NBR with 34% acrylonitrile content (Chapter 5). This further demonstrates the sensitivity of crosslink distribution to even rather small changes in relative polarity of the rubbers in a blend. The addition of MG30 appears to reduce the maldistribution in crosslink density, presumably as a result of reduced phase sizes and the greater role of migration of intermediates during vulcanization.

## 6.6 CONCLUSIONS

It has been shown that the addition of a graft copolymer of NR, Heveaplus MG30, to NR/NBR blends can reduce phase sizes attained in the blends over a wide range of acrylonitrile content of the NBR. The addition of MG30 is beneficial to tensile strength. Whilst this may simply be a consequence of the smaller phase sizes, it also suggests an improved integrity of the vulcanized blend and hence improved interfacial adhesion.

The addition of MG30 alters the distribution of crosslinks between the phases of the blends. This may be due to the reduction in phase size permitting migration of vulcanization intermediates by diffusion during vulcanization to play a more important role in determining the final crosslink distribution. It suggests that optimization of the cure system should be made in the compatibilized blend if it is intended to use compatibilizing agents. A cure system shown to be appropriate or even ideal in the binary blend may no longer be suitable.

## 6.7 REFERENCES

1 Setua, D.K. and White, J.L. (1991) Flow visualization of the mixing and blend phase morphology of binary and ternary blends of polychloroprene, acrylonitrile-butadiene-copolymer and ethylene-propylene-rubber. *Kautschuk Gummi Kunststoffe*, **44**(9), 821–826.

2 Gaylord, N.G. (1975) in *Copolymers, polyblends and composites*, ed. A.J. Platzer. Washington, DC, American Chemical Society.

3 Locke, C.E. and Paul, D.R. (1973) Graft copolymer modification of polyethylene-polystyrene blends. *Journal of Applied Polymer Science*, **17**, 2597–2617; 2791–2800.

4 Barentsen, W.M. and Heikens, D. (1973) Mechanical properties of polystyrene/low density polyethylene blends. *Polymer*, **14**(11), 579–583.

5 Barentsen, W.M., Heikens, D. and Piet, P. (1974) Effect of addition of graft copolymer on the microstructure and impact strength of PS/LDPE blends. *Polymer*, **15**(2), 119–122.

6 Ide, F. and Hasegawa, A. (1974) Study of polymer blend of nylon 6 and polypropylene or nylon 6 and polystyrene using the reaction of polymer. *Journal of Applied Polymer Science*, **18**, 963–974.

7 Avgeropoulos, G.N., Weissert, F.C., Biddison, P.H. and Böhm, G.G.A. (1976) Heterogeneous blends of polymers. Rheology and morphology. *Rubber Chemistry and Technology*, **49**(1), 93–104.

8 Hess, W.M., Scott, C.E. and Callan, J.E. (1967) Carbon black distribution in elastomer blends, *Rubber Chemistry and Technology*, **40**(2), 371–384.

9 Folt, V.L. and Smith, R.W. (1973) Rheology of elastomer blends. *Rubber Chemistry and Technology*, **46**(5), 1193–1209.

10 Andrews, E.H. and Turner, D.T. (1960) The distribution of polymethyl methacrylate formed in natural rubber latex: an electron microscopical study. *Journal of Applied Polymer Science*, **3**(9), 366–367.

11 Lewan, M.V. (1996) *Crosslink density distribution in NR/NBR blends*. PhD Thesis, Loughborough University, p. 119.

# 7

# Improving the morphology and properties of NR/NBR blends with polychloroprene as the compatibilizing agent

*Krisna Kongsin*[*] *and Michael V. Lewan*[†]
[*]*Rubber Research Institute of Thailand, Thailand;*
[†]*Tun Abdul Razak Research Centre, UK*

## 7.1 INTRODUCTION

Extensive work on compatibilizing NR/NBR blends with Heveaplus MG30 had shown that this graft copolymer of natural rubber with poly(methyl methacrylate) provided improved adhesion between polymer phases with incompatible polarities, which was crucial to stress transfer between these phases (Chapter 6). In addition, the interfacial tension between the phases was lowered by this interfacial activity, and this played an important role in blend morphology, reducing phase sizes and subsequently improving physical properties and crosslink density distribution between the phases of the blend.

Potential applications for NR/NBR blends include food contact applications which are subject to restrictions in materials use in order to comply with the statutory requirements of regulatory bodies such as the US Food and Drugs Administration (FDA). Unfortunately, Heveaplus MG is not approved by the FDA whereas polychloroprene is. Polychloroprene (CR) has a similar backbone structure to NR but is more polar due to the chlorine substituent. Gum compounds, or those with

*Blends of Natural Rubber*. Edited by Andrew J. Tinker and Kevin P. Jones.
Published in 1998 by Chapman & Hall, London. ISBN 0 412 819406

non-reinforcing filler, have high mechanical strength, and because of the polarity, a good resistance to swelling by aliphatic oils [1]. Importantly for use in the context considered here, CR has a solubility parameter intermediate between those of NR and NBR. Furthermore, the presence of a dipole within the repeat unit allows the possibility of dipolar interaction with the acrylonitrile repeat unit of NBR. These factors suggest that CR may act as a compatibilizing agent for NR/NBR blends by locating preferentially at the NR/NBR interface. Some evidence for this has already been recorded [2].

## 7.2 CURE SYSTEMS FOR NR/CR/NBR BLENDS

### 7.2.1 Materials

NBR is available in a range of acrylonitrile contents, the higher the acrylonitrile content the higher the resistance to swelling by oils. Grades containing medium to high acrylonitrile levels, about 34 to 41%, are most often considered for blends with NR. NBR at the former end of the compositional range will be referred to as NBR34 and at the latter as NBR41. The two specific grades of NBR used were Krynac 34.50 (Bayer) and Breon N41C45 (Zeon Chemicals). The polychloroprene used was Neoprene WRT (DuPont). Only the most promising cure systems identified in earlier work on crosslink distributions in NR/NBR blends (Chapter 5) were included in the investigation of CR as a compatibilizing agent.

Curing systems for 50:50 NR/NBR34 blends were investigated, with the substitution of 10 phr NR by CR, and also the substitution of 5, 10 and 20 phr NBR by CR. Masterbatches based on NR and NBR34 (Table 7.1) were prepared in an internal mixer and then the masterbatches were blended together on a two-roll mill. The addition of curatives was carried out at 30 °C according to the amounts shown in Table 7.2. Test sheets 2 mm thick were compression moulded and vulcanized in a press at 180 °C for the time to maximum torque rise ($t_{max}$) as estimated from rheographs obtained using a Monsanto MDR 2000 Rheometer. This cure

**Table 7.1** Masterbatch formulations (phr)

| | | | | | | |
|---|---|---|---|---|---|---|
| SMR L | 100 | 80 | – | – | – | – |
| Krynac 34.50 | – | – | 100 | 90 | 80 | 60 |
| CR | – | 20 | 2 | 10 | 20 | 40 |
| Zinc oxide | 5 | 5 | 5 | 5 | 5 | 5 |
| Stearic acid | 2 | 2 | 2 | 2 | 2 | 2 |
| TMQ | 1.5 | 1.5 | 1.5 | 1.5 | 1.5 | 1.5 |

**Table 7.2** Cure systems for NR/CR/NBR blends

| *S (phr)* | *CBS (phr)* | *MBS (phr)* | *TBBS (phr)* | *TBTD (phr)* | *TETD (phr)* |
|---|---|---|---|---|---|
| 1.3 | 1.3 | – | – | – | – |
| 1.3 | 1.3 | – | – | 0.2 | – |
| 1.3 | 1.3 | – | – | – | 0.14 |
| 1.3 | – | 1.24 | – | – | – |
| 1.3 | – | 1.24 | – | 0.2 | – |
| 1.3 | – | 1.24 | – | – | 0.14 |
| 1.3 | – | – | 1.17 | – | – |
| 1.3 | – | – | 1.17 | 0.2 | – |
| 1.3 | – | – | 1.17 | – | 0.14 |

temperature was chosen as being representative of processing by injection moulding, since this is a common production technique for envisaged potential applications.

Crosslink densities were estimated by swollen-state $^1H$ NMR spectroscopy on a Bruker AC300 spectrometer as described in Chapter 2. The phase morphology of NR/NBR blends was investigated by STEM microscopy (Chapter 3).

### 7.2.2 Effect of polychloroprene content on crosslink distribution and physical properties

The tensile strengths of gum 50:50 NR/NBR34 blends are quite low (Table 7.3), with the exception of blends cured with CBS/TBTD or CBS/

**Table 7.3** Physical properties of 50:50 NR:NBR blends

| *Accelerator system* | *M100 (MPa)* | *M300 (MPa)* | *Tensile strength (MPa)* | *Elongation at break (%)* |
|---|---|---|---|---|
| CBS | 0.66 | 1.08 | 8.6 | 745 |
| CBS/TBTD | 0.76 | 1.26 | 18.8 | 745 |
| CBS/TETD | 0.70 | 1.14 | 16.1 | 760 |
| MBS | 0.68 | 1.06 | 11.7 | 815 |
| MBS/TBTD | 0.74 | 1.26 | 9.9 | 695 |
| MBS/TETD | 0.74 | 1.21 | 11.5 | 690 |
| TBBS | 0.64 | 1.02 | 9.3 | 780 |
| TBBS/TBTD | 0.71 | 1.15 | 10.8 | 745 |
| TBBS/TETD | 0.73 | 1.21 | 13.3 | 720 |

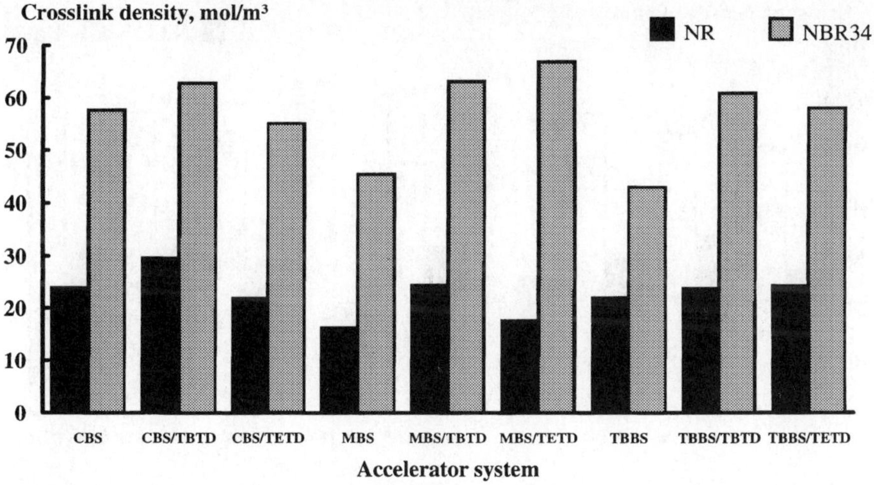


**Figure 7.1** Crosslink density distributions in 50:50 NR:NBR34 gum blends.

TETD. The crosslink density distribution between the two rubber phases favours the NBR phase for all of the cure systems (Figure 7.1).

Initial investigation of CR as compatibilizer for this blend system considered moderate levels of CR, 10–20 phr. Substitution of 10 phr of the NR in the blend with 10 phr CR increases the tensile strength of the blends (Table 7.4). This may in part be a consequence of a concomitant change in the crosslink distribution. Whilst the NBR still has the higher crosslink density, the bias in crosslink density is reduced mainly as a result of increased crosslinking in the NR phase of the blend (Figure 7.2).

**Table 7.4** Physical properties of 40:10:50 NR:CR:NBR blends

| *Accelerator system* | *M100 (MPa)* | *M300 (MPa)* | *Tensile strength (MPa)* | *Elongation at break (%)* | *Crescent tear strength (N/mm)* |
|---|---|---|---|---|---|
| CBS | 0.71 | 1.32 | 17.0 | 785 | 30.3 |
| CBS/TBTD | 0.74 | 1.39 | 16.9 | 740 | 19.1 |
| CBS/TETD | 0.73 | 1.41 | 17.7 | 735 | 15.3 |
| MBS | 0.71 | 1.30 | 16.7 | 785 | 15.1 |
| MBS/TBTD | 0.75 | 1.45 | 14.7 | 690 | 13.4 |
| MBS/TETD | 0.76 | 1.52 | 14.0 | 650 | 10.3 |
| TBBS | 0.71 | 1.31 | 18.5 | 800 | 21.1 |
| TBBS/TBTD | 0.74 | 1.43 | 15.2 | 725 | 17.8 |
| TBBS/TETD | 0.73 | 1.40 | 14.6 | 705 | 15.7 |

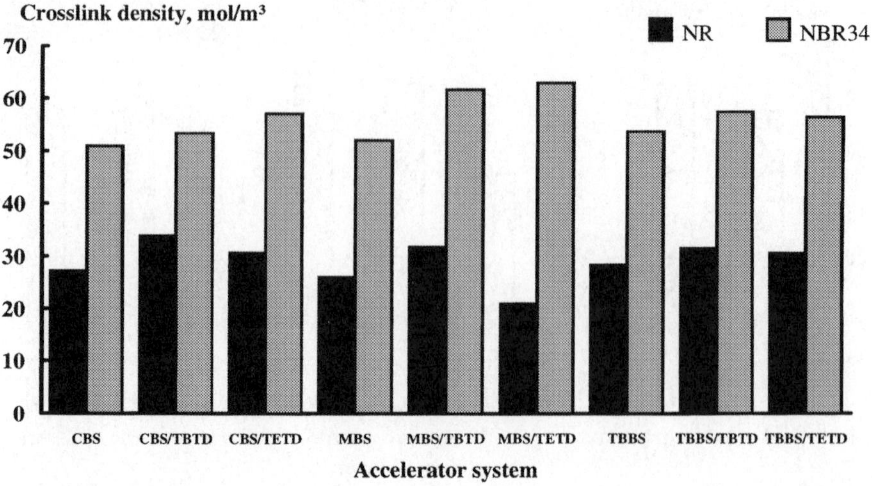


**Figure 7.2** Crosslink density distributions in 40:10:50 NR:CR:NBR34 gum blends.

If 10 phr NBR34 is replaced with CR, there is also a reduction in the maldistribution of crosslink density in favour of the NBR, although to a lesser extent (Figure 7.3). On the other hand, the tensile strengths of the blends are higher (Table 7.5). This may be attributed to the higher content of NR, which has inherently high strength.

Increasing the substitution of NBR34 by CR to 20 phr further reduces the basis of crosslink density in favour of NBR (Figure 7.4). This is

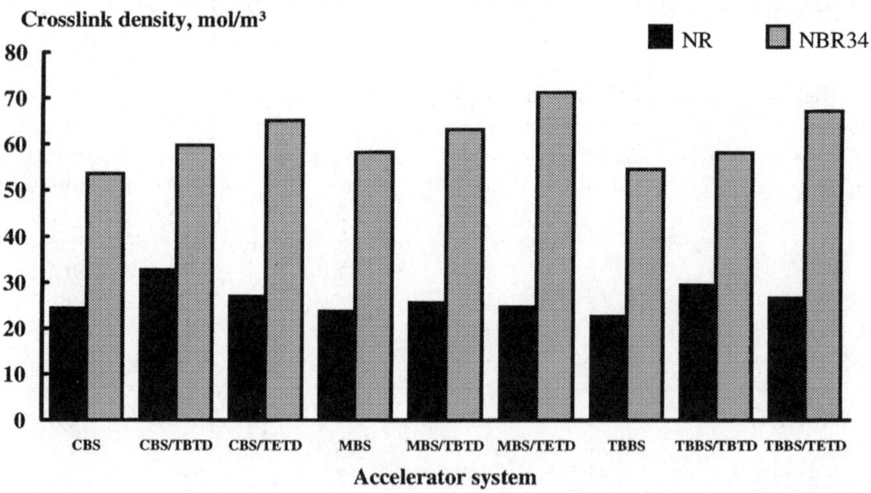


**Figure 7.3** Crosslink density distributions in 50:10:40 NR:CR:NBR34 gum blends.

**Table 7.5** Physical properties of 50:10:40 NR:CR:NBR blends

| *Accelerator system* | *M100 (MPa)* | *M300 (MPa)* | *Tensile strength (MPa)* | *Elongation at break (%)* | *Crescent tear strength (N/mm)* |
|---|---|---|---|---|---|
| CBS | 0.68 | 1.26 | 19.5 | 810 | 26.6 |
| CBS/TBTD | 0.75 | 1.48 | 19.8 | 730 | 24.9 |
| CBS/TETD | 0.73 | 1.44 | 18.8 | 725 | 14.9 |
| MBS | 0.69 | 1.29 | 18.5 | 785 | 22.8 |
| MBS/TBTD | 0.72 | 1.42 | 18.7 | 725 | 19.3 |
| MBS/TETD | 0.73 | 1.46 | 17.0 | 685 | 14.6 |
| TBBS | 0.65 | 1.17 | 18.1 | 810 | 23.5 |
| TBBS/TBTD | 0.71 | 1.37 | 18.2 | 740 | 29.2 |
| TBBS/TETD | 0.75 | 1.54 | 20.7 | 740 | 14.0 |

generally achieved by a combination of increased crosslinking in the NR and reduced crosslinking in the NBR. However, increases in tensile strength are small (Table 7.6); indeed, tear strength is reduced.

CR is clearly efficacious in terms of improving tensile strength, and the cure systems based on CBS generally give the best ultimate properties. The effect of substitution of just 5 phr of NBR34 with CR was therefore considered only in vulcanizates cured with the three systems based on CBS.

The tensile strengths of these blends were the highest recorded (Table 7.7), with the exception of the 50:20:30 NR:CR:NBR34 blend cured

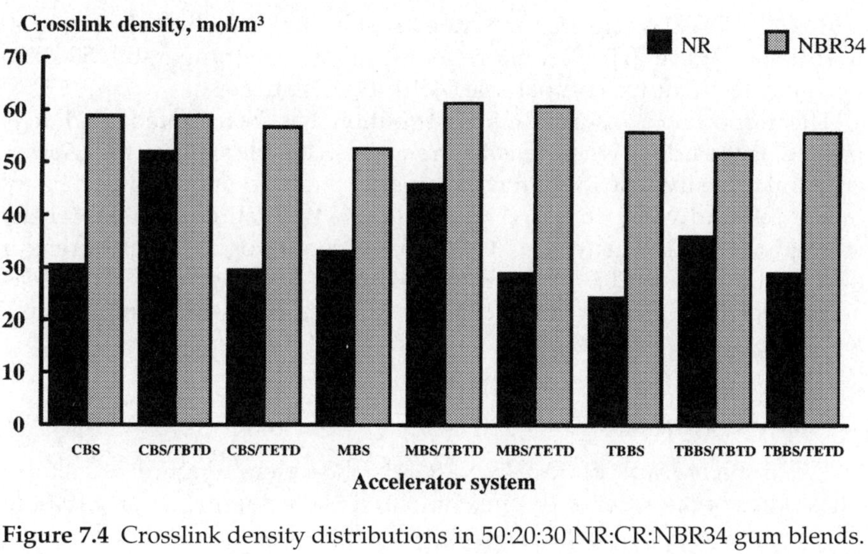


**Figure 7.4** Crosslink density distributions in 50:20:30 NR:CR:NBR34 gum blends.

**Table 7.6** Physical properties of 50:20:30 NR:CR:NBR blends

| *Accelerator system* | *M100 (MPa)* | *M300 (MPa)* | *Tensile strength (MPa)* | *Elongation at break (%)* | *Crescent tear strength (N/mm)* |
|---|---|---|---|---|---|
| CBS | 0.71 | 1.47 | 18.5 | 790 | 14.5 |
| CBS/TBTD | 0.83 | 1.78 | 21.5 | 760 | 12.7 |
| CBS/TETD | 0.69 | 1.39 | 19.5 | 765 | 14.2 |
| MBS | 0.65 | 1.36 | 17.8 | 800 | 10.1 |
| MBS/TBTD | 0.80 | 1.79 | 21.6 | 725 | 12.3 |
| MBS/TETD | 0.71 | 1.55 | 19.6 | 760 | 13.9 |
| TBBS | 0.65 | 1.39 | 18.8 | 810 | 10.4 |
| TBBS/TBTD | 0.68 | 1.45 | 20.0 | 790 | 14.9 |
| TBBS/TETD | 0.70 | 1.55 | 17.7 | 755 | 11.8 |

**Table 7.7** Physical properties of 50:5:45 NR:CR:NBR blends

| *Accelerator system* | *M100 (MPa)* | *M300 (MPa)* | *Tensile strength (MPa)* | *Elongation at break (%)* | *Crescent tear strength (N/mm)* |
|---|---|---|---|---|---|
| CBS | 0.69 | 1.27 | 20.8 | 800 | 25.6 |
| CBS/TBTD | 0.73 | 1.40 | 21.0 | 750 | 12.9 |
| CBS/TETD | 0.74 | 1.45 | 22.8 | 760 | 19.7 |

with CBS/TBTD as accelerator. Remarkably, the crosslink density distributions (Figure 7.5) were more akin to those for comparable 50:20:30 vulcanizates than for comparable 50:10:40 vulcanizates.

The importance of crosslink distribution has been noted, and it is particularly evident when tensile strength is considered as a function of crosslink density distribution expressed as the ratio of crosslink density in the NR and NBR phases (Figure 7.6). As this ratio increases, tensile strength generally increases. This may account for the high strength observed for the 50:20:30 NR:CR:NBR34 blend cured with CBS/TBTD as accelerator. Explanation of the changes in crosslink distribution requires an understanding of the phase morphology of the blends.

## 7.3 PHASE MORPHOLOGY OF NR/CR/NBR BLENDS

A STEM micrograph of 50:5:45 NR:CR:NBR34 blend (Figure 7.7) shows phase sizes of the order of 10 μm (compared to the phase sizes of >100 μm for the non-compatibilized blends); these have been generated from the

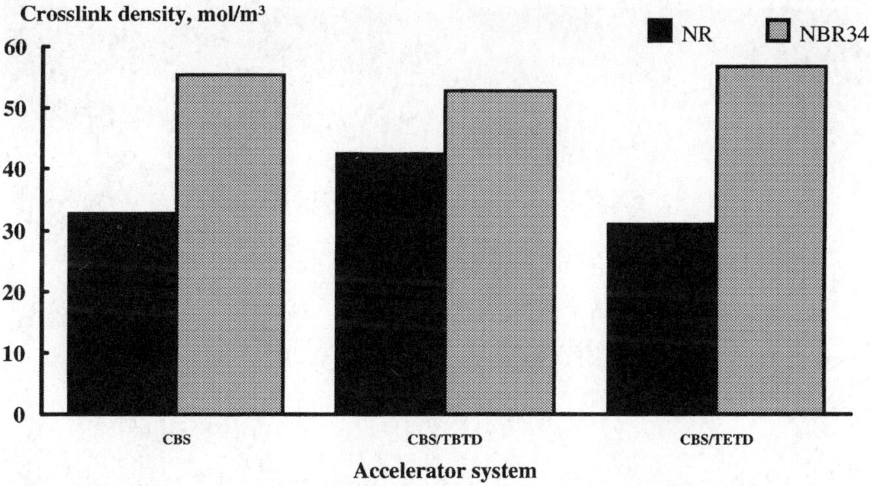


**Figure 7.5** Crosslink density distributions in 50:5:45 NR:CR:NBR34 gum blends.

reduction in interfacial tension at the NR/NBR interface. The NR phase is more heavily stained with osmium tetroxide than the NBR phase and so is the darker component in the micrograph. Small pockets of NR are also visible within the NBR phases and vice versa, while the black spots of zinc oxide are present only in the NR phase.

A thin black line is evident at the interface between the NR and NBR phases. This is the polychloroprene, which appears dark in the micro-

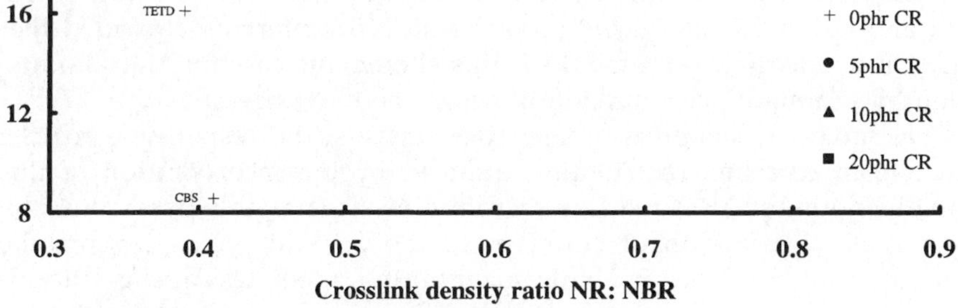


**Figure 7.6** Dependence of tensile strength in the ratio of crosslink densities in the NR and NBR phase of NR/CR/NBR34 blends vulcanized at 180 °C.

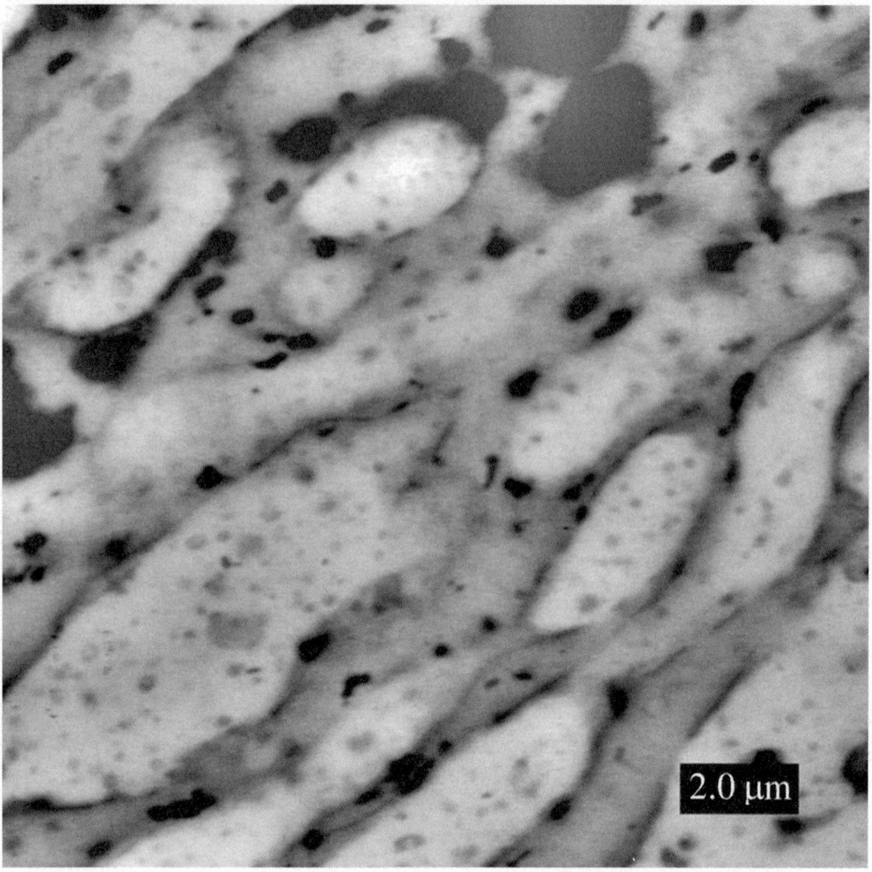


**Figure 7.7** STEM micrograph of a 50:5:45 NR:CR:NBR34 gum blend.

graph due to its chlorine content. The presence of the polychloroprene at the interface reduces interfacial tension and hence the phase size attained in the blend. A greater mixing between elastomers is expected at the NR/CR and CR/NBR interface than at a NR/NBR interface. Provided that the CR is adequately crosslinked, this should increase interfacial adhesion by promoting crosslinking between the two phases.

The marked change in phase size may also be responsible for the change in crosslink distribution induced by the incorporation of this small amount of CR. Crosslink distributions are not only determined by the initial distribution of curatives at the onset of vulcanization; the migration of curatives and vulcanization intermediates during vulcanization can also play a role, provided phase sizes are sufficiently small. If phase sizes are large, as found in NR/NBR blends in the absence of compatibilizing agents, migration during vulcanization can only affect crosslinking in the vicinity of the interface and hence will play a small part in determining the overall crosslink distribution. However, when phase sizes are reduced by a compatibilizing agent, migration of vulcanization intermediates can play a significant role in determining overall crosslink distributions.

40:10:50 and 50:10:40 NR:CR:NBR34 blends have similar phase morphologies (Figures 7.8 and 7.9) and slightly smaller phase sizes than observed for the blend containing 5 phr CR. There is a further reduction in phase size at the higher CR content of 50:20:30 NR:CR:NBR34 blends (Figure 7.10), but the phase morphology also becomes more complex. There appear to be strands of polychloroprene throughout the NBR phase; these could either be simply in the NBR phase or at the boundary of NR microdomains within the NBR phase. There are also thin strands,

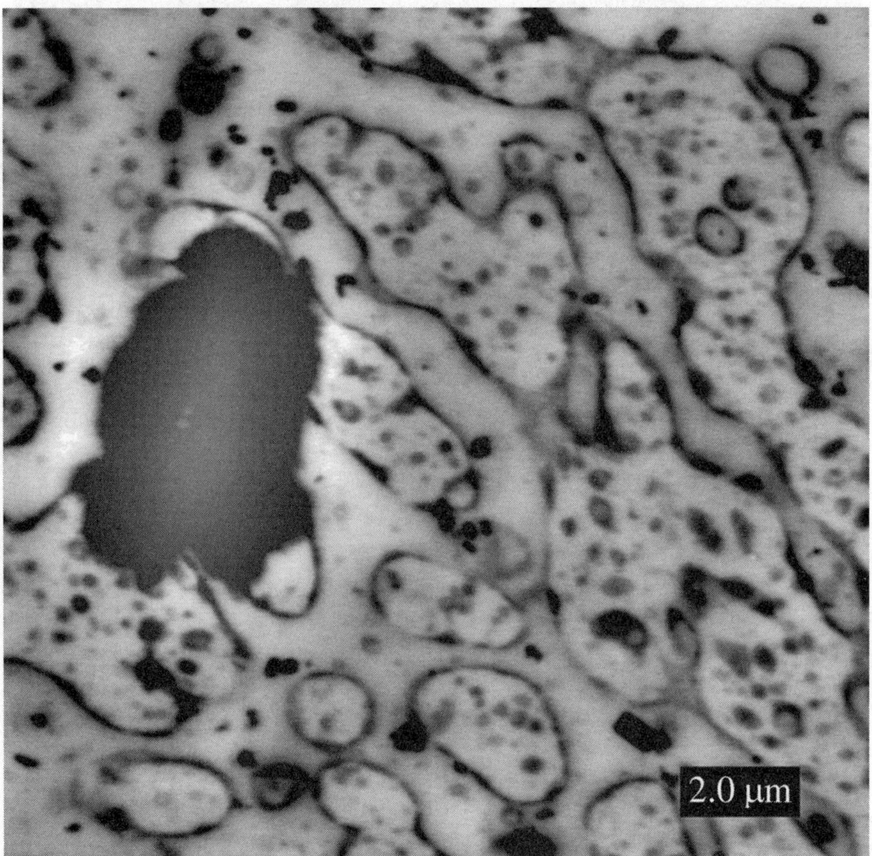


**Figure 7.8** STEM micrograph of a 40:10:50 NR:CR:NBR34 gum blend.

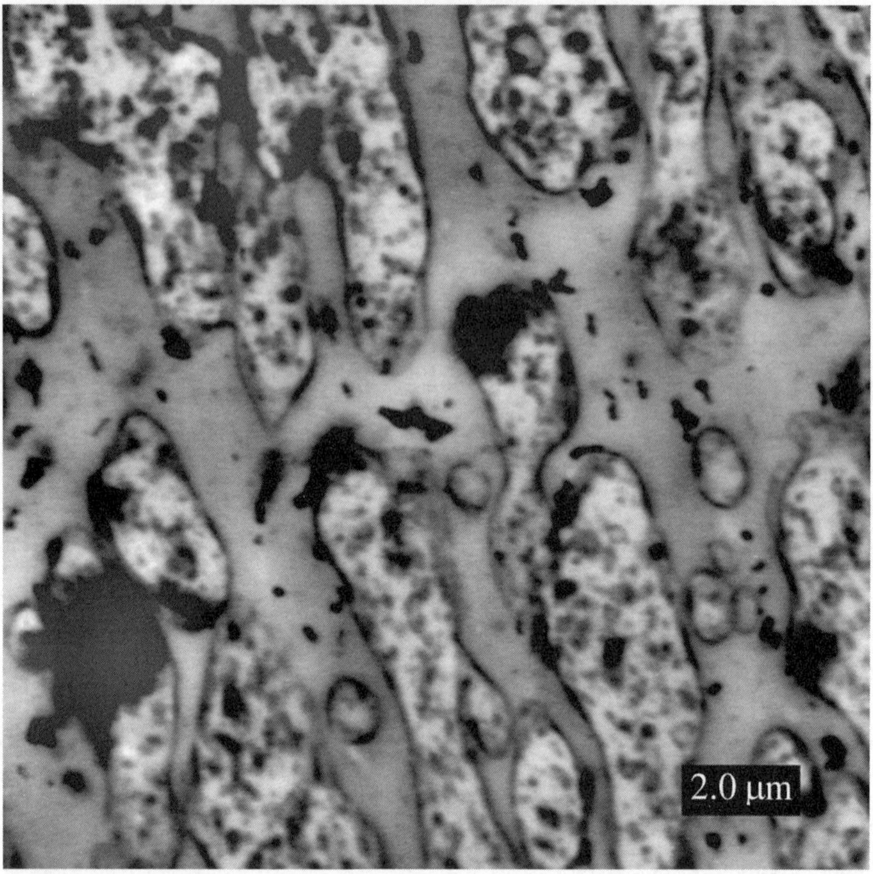


**Figure 7.9** STEM micrograph of a 50:10:40 NR:CR:NBR34 gum blend.

which could be polychloroprene, going across the NR phase and linking up with other NBR phases.

There is clearly little benefit in using much more than 5 phr of CR as compatibilizer. The morphological studies do not provide an explanation of the generally lower NR:NBR crosslink density ratios at the higher CR contents. Total crosslink densities measured in the two major phases of the blends do not suggest that there is significant loss of crosslink to the CR component. Indeed, there could be concern regarding adequacy of crosslinking of the CR, and this could be a factor in the generally lower strength of these blends.

CR is also effective in reducing phase sizes in NR/NBR41 blends, as may be seen by comparing Figure 7.11 with Figure 6.1a of Chapter 6. Changes in crosslink distribution and tensile strength analogous to those observed for NR/CR/NBR34 blends may also be seen (Figure 7.12).

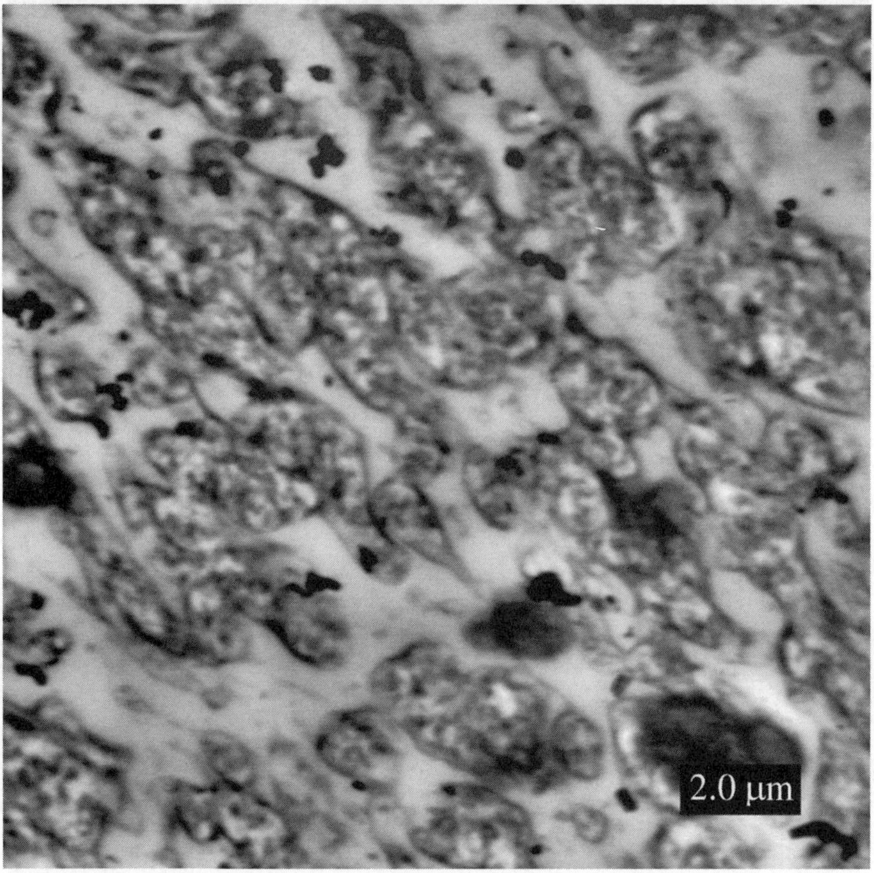


**Figure 7.10** STEM micrograph of a 50:20:30 NR:CR:NBR34 gum blend.

Small phase sizes are also seen in practical compounds derived from these blends. This is illustrated by the TEM micrograph of a 50:5:45 NR:CR:NBR34 blend containing 20 phr N660 carbon black and 40 phr calcium carbonate depicted in Figure 7.13.

## 7.4 CONCLUSIONS

Polychloroprene is a very good compatibilizer for NR/NBR blends. To obtain the best tensile strengths 5 phr polychloroprene is substituted for 5 phr NBR. This provides a good reduction in phase sizes and reduces the pronounced maldistribution of crosslinks in favour of the NBR phase in blends with NBR34. Addition of more polychloroprene will reduce the phase sizes further, but does not offer further improvements in tensile strength. For blends based on NBR containing 34% acrylonitrile, cure

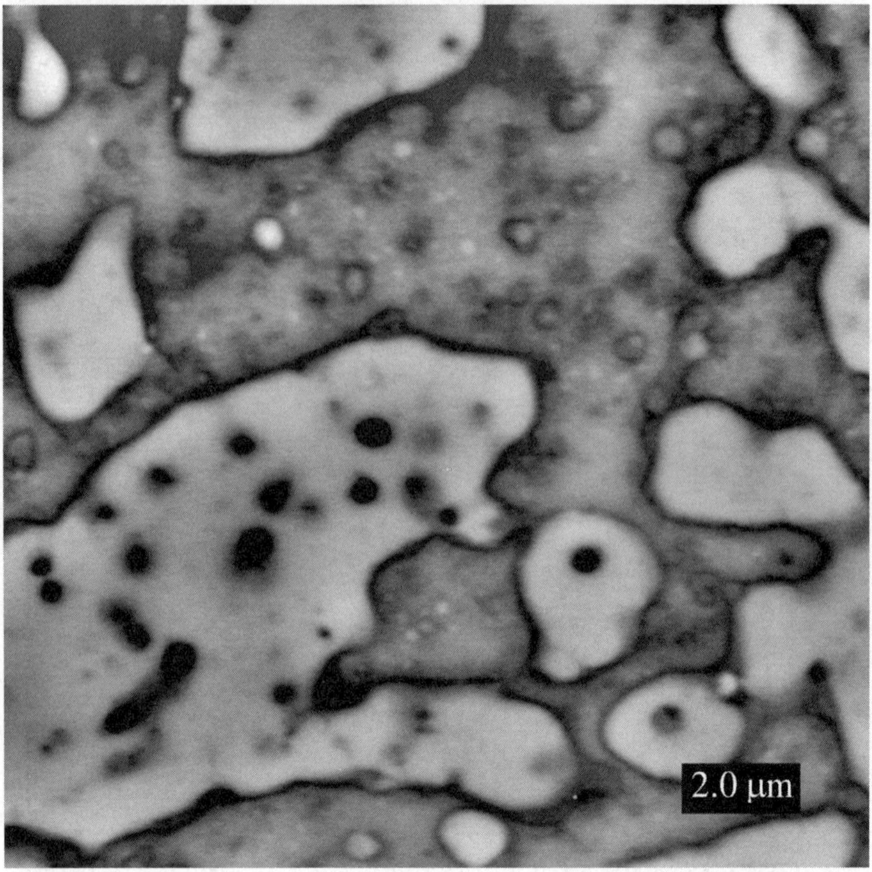


Figure 7.11 STEM micrograph of a 40:10:50 NR:CR:NBR41 gum blend.

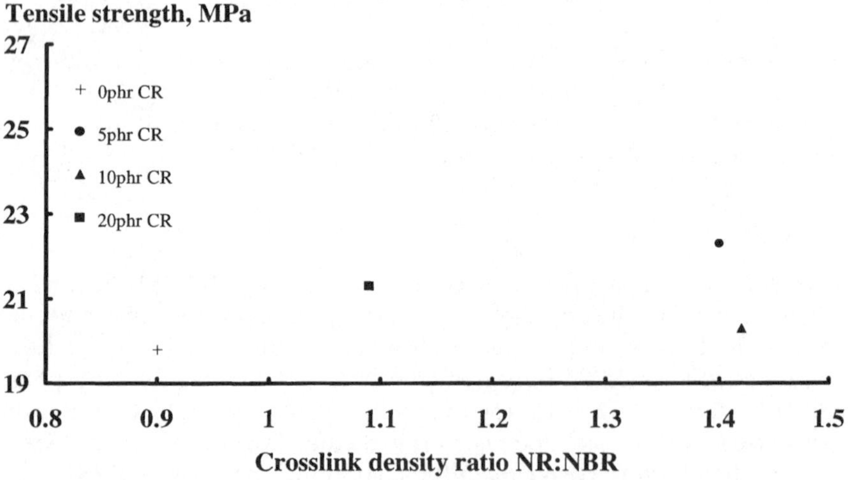


Figure 7.12 Dependence of tensile strength on the ratio of crosslink densities in the NR and NBR phases of NR/CR/NBR41 blend vulcanizates.

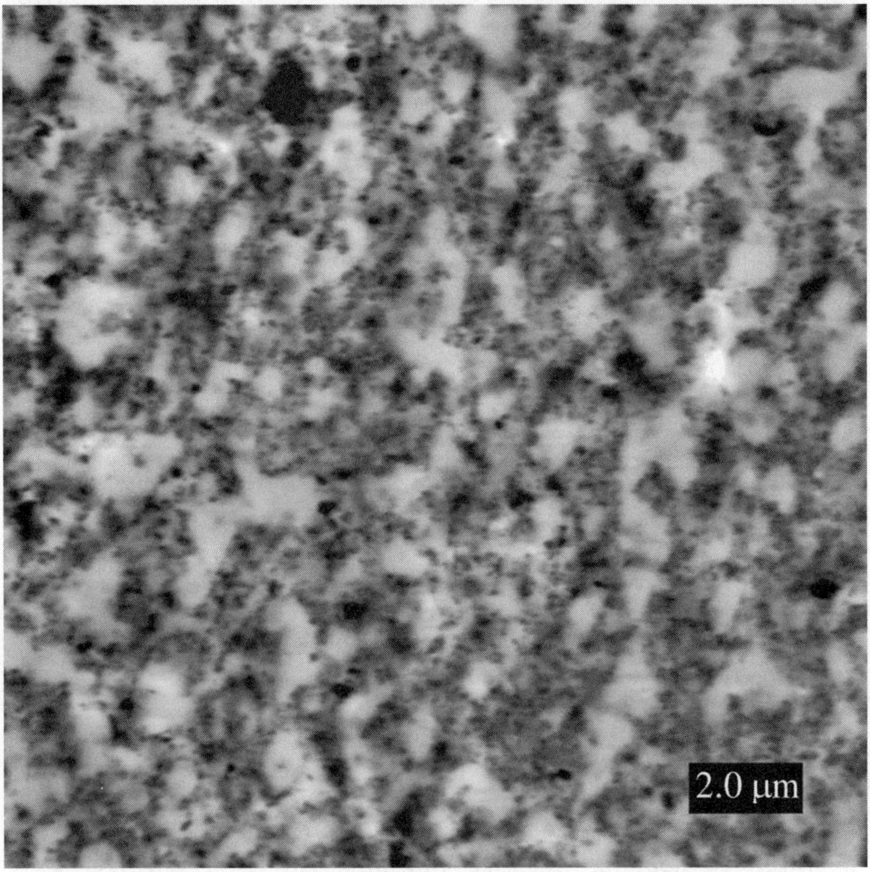


**Figure 7.13** TEM micrograph of a 50:5:45 NR:CR:NBR34 blend filled with carbon black (20 phr N660) and calcium carbonate (40 phr).

systems containing CBS, particularly accelerator combinations of CBS/TBTD and CBS/TETD, are recommended.

## 7.5 REFERENCES


1 Murray, R.M. and Thomson, D.C. (1964) *The Neoprenes* (International Edition) Wilmington, Delaware: du Pont.

2 Tinker, A.J. (1987) Blends of natural rubber with speciality synthetic rubbers, in *Industrial Composites based on Natural Rubber*, Malaysian Rubber Research and Development Board, Kuala Lumpur, pp. 103–113.

8

# NR/NBR blends – compounding for food contact applications

*Michael V. Lewan*
*Tun Abdul Razak Research Centre, MRPRA, UK*

## 8.1 INTRODUCTION

The initial work into preferred formulations for NR/NBR blends produced a series of cure systems suitable for use in blends of NR with NBR containing either 41% or 34% acrylonitrile (Chapter 5). The blends may be further improved by the use of low levels of compatibilizers, either polychloroprene (CR) or a graft copolymer of natural rubber with poly(methyl methacrylate) (Heveaplus MG30) as described in Chapter 6.

As an example of a practical use of these blends, formulations suitable for use in the manufacture of milking inflations have been developed. Milking inflations are used in the dairy industry and, as they are the part of the milking machine which fits on the teat of a cow (Figure 8.1), they come into direct contact with milk. Therefore, all ingredients must conform to regulations governing food contact. These differ from country to country, but those set by the US Food and Drugs Administration (FDA) [1] are particularly pertinent because of both the size of the US market and the use of these regulations elsewhere.

The permitted additives for milking inflations come under Section 177.2600: Rubber articles intended for repeated use. A wide range of fillers is allowed, but a limit of 10% by weight of the total formulation is imposed on furnace carbon black. This limitation can prove very difficult for NBR, which lacks the inherent high strength of NR, because it is difficult to obtain adequate physical properties for certain designs of

*Blends of Natural Rubber*. Edited by Andrew J. Tinker and Kevin P. Jones.
Published in 1998 by Chapman & Hall, London. ISBN 0 412 819406

**Figure 8.1** A milking inflation.

milking inflation. A wide range of plasticizers is also tolerated, but the total must not exceed 30% w/w. The selection of N660 black and dioctyl phthalate (DOP) as plasticizer in the work considered here is consistent with common practice for milking inflations. Two white fillers were used; silica was initially selected but, as this is relatively expensive, cheaper calcium carbonate was also investigated.

## 8.2 FORMULATIONS FOR MILKING INFLATIONS TO MEET FDA REGULATIONS

### 8.2.1 Compounding to meet a specification

Milking inflations have to meet certain specifications set by suppliers to the dairy industry. One particular specification is presented in Table 8.1. In addition to the specification, practical considerations dictate that a higher tear strength would be desirable, if not necessary. Substantial hot tear resistance is required in order to ensure de-moulding from the injection-mould cavity without tearing of the inflation.

Resistance to swelling by butter oil is a key requirement of milking inflations. For a range of nitrile rubbers differing in acrylonitrile content, it is seen that those containing either 41 wt% acrylonitrile, NBR41, as typified by Breon N41 (Zeon Chemicals) or 34 wt% acrylonitrile, NBR34, as typified by Krynac 34.50 (Bayer) swell very little in butter oil (Figure 8.2) while NR is known to have very poor resistance to swelling in oils.

The carbon black was placed wholly within the NR masterbatch at a loading of 25 phr in order to make best use of restriction of swelling arising from rubber–filler interaction. The carbon black is not required to reinforce the NBR phase – the NR phase will impart the necessary

**Table 8.1** Typical milking inflation specification (USA)

| | |
|---|---|
| **Unaged properties** | |
| Tensile strength | 10.5 MPa (min) |
| Elongation at break | 350% (min) |
| Crescent tear strength | 15 N/mm (min) |
| Hardness | 50 Shore A° (+/−3) |
| Elongation after 90 min at 1 N/mm$^2$ | 60 to 100% |
| Subsequent tension set | 2.5% (max) |
| **After heat ageing (72 hours at 100 °C)** | |
| Change in tensile strength | 20% (max) |
| Change in elongation at break | 50% (max) |
| Change in hardness | 20 Shore A° (max) |
| Change in elongation at 1 N/mm$^2$ | 35% (max) |
| **Resistance to liquids (168 hours at 70 °C)** | |
| (a) *Butter oil* | |
| Volume change | ±10% |
| Hardness change | +6/−5 Shore A° |
| (b) *Distilled water* | |
| Volume change | +5/−10% |
| Hardness change | + 5/−6 Shore A° |

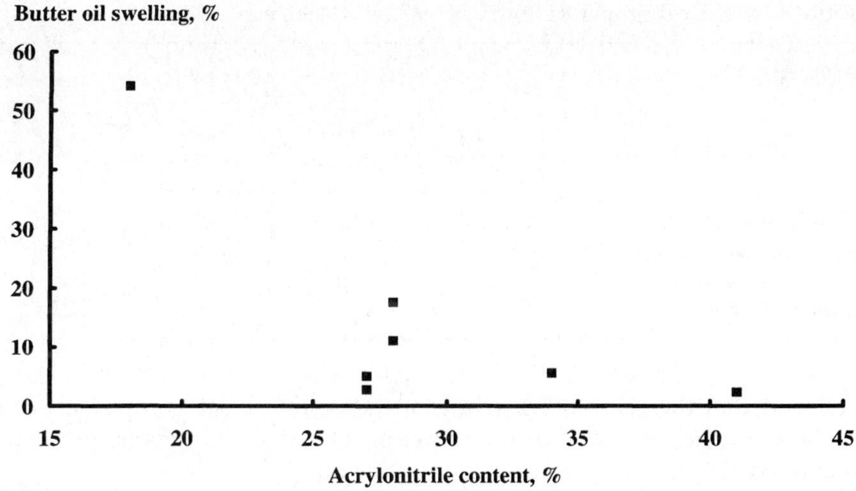


**Figure 8.2** Dependence of volume swelling of NBR vulcanizates in butter oil on acrylonitrile content.

strength properties to the blend (Chapter 5). Initially, the NBR masterbatch was filled with a low level of white filler, silica (10 phr). The plasticizer was distributed between the two masterbatches to ensure that the fillers mixed effectively. In addition to filler and plasticizer, both masterbatches contained 5 phr zinc oxide, 2 phr stearic acid and 3 phr of the antidegradant IPPD. A number of promising cure systems identified in investigations of crosslink distribution in gum vulcanizates (Chapter 5) were used; all contained 1.3 phr sulphur. Separate masterbatches of the NR and the NBR were each prepared in an internal mixer, and then cross-blended to give a 50:50 polymer ratio and finalized with curatives on a two-roll mill. The blends were compression moulded to the optimum cure time at 180 °C.

Limited physical testing on both blends demonstrated that whilst tensile strengths were high, resistance to swelling in butter oil was poor. Unaged physical properties of 50:50 NR:NBR41 blends cured using the preferred accelerator systems are shown in Table 8.2. 50:50 NR:NBR34 blends had very similar properties. Crescent tear strengths are very good, hardness values are slightly low but butter oil swellings exceed the permitted FDA maximum by about six times. The crosslink density distribution between the two rubber phases was undesirable for both the NBR41 and NBR34 blends, with both systems showing a marked bias in favour of the NBR phase (Figures 8.3 and 8.4).

Despite the level of crosslinking in the NR phase of the NR/NBR41 blends being very low, the blends had fairly high tensile strengths. It was

**Table 8.2** Physical properties for 50:50 NR:NBR41 blends

| *Property* | *Accelerator system (phr)* | | | |
|---|---|---|---|---|
| | *MBS/MBTS (1.24/0.1)* | *MBS/TBTD (1.24/0.18)* | *TBBS (1.17)* | *TBBS/DPG (1.17/0.1)* |
| Tensile strength, MPa | 17.8 | 17.8 | 19.2 | 15.7 |
| Crescent tear, N/mm | 30.1 | 41.7 | 35.0 | 27.4 |
| Hardness, Shore A | 41.5 | 45.5 | 42.5 | 42.5 |
| Butter oil swelling, % | 63.7 | 61.7 | 67.6 | 52.3 |

believed that the reinforcing action of the carbon black compensates to some extent for the lack of crosslinking. However, the major problem remained the very high swelling in butter oil.

Replacement of the 10 phr silica used in the NBR masterbatch with 60 phr of an alternative – and cheaper – white filler, calcium carbonate, was investigated as an indirect means of reducing the level of swelling in butter oil. Addition of a relatively high level of inert white filler to the NBR phase, and the corresponding increase in DOP plasticizer (20 phr) to maintain hardness, serves to increase the formula weight of the blend compound. This enables the level of carbon black in the NR phase to be increased from 25 to 35 phr, whilst still remaining within the FDA permitted limit of 10 wt% carbon black overall. As seen in Table 8.3, the addition of a high level of inert filler, albeit only to the NBR phase, does

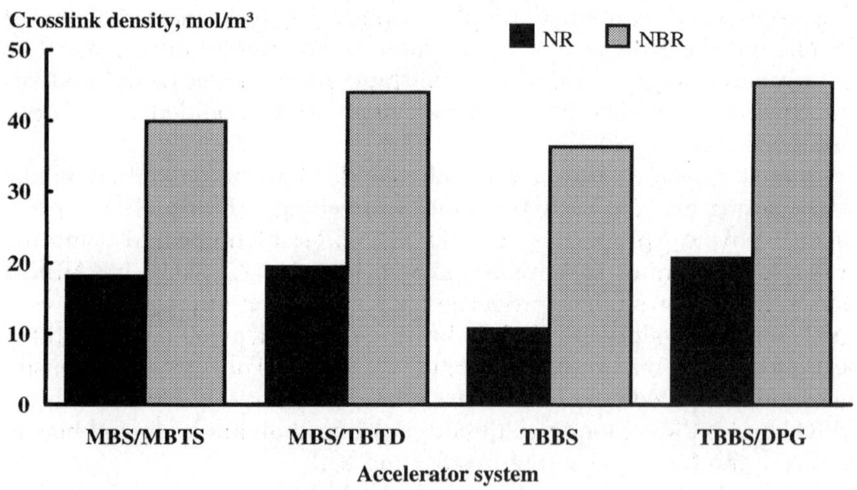


**Figure 8.3** Crosslink density distributions in 50:50 NR:NBR41 blends.

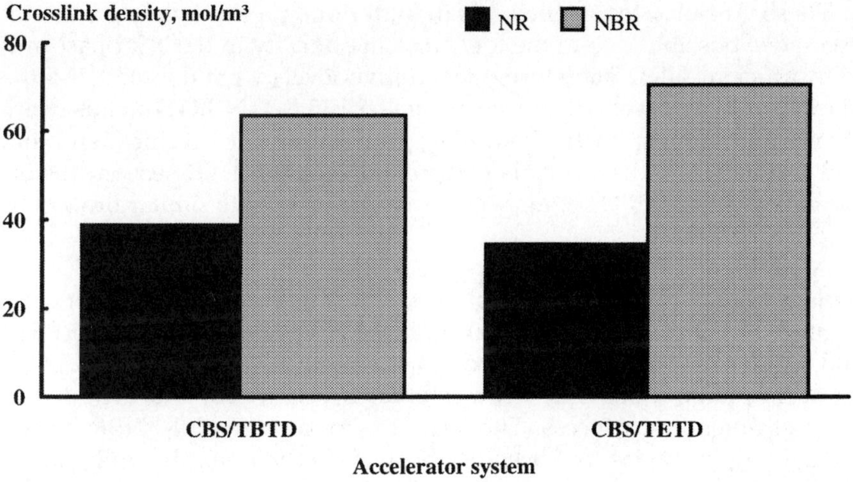


**Figure 8.4** Crosslink density distributions in 50:50 NR:NBR34 blends.

reduce the tensile strength of the blend, although it remains above the minimum specification. A significant reduction of swelling in butter oil is also achieved, although the levels shown in Table 8.3 remain too high.

Some different distributions of filler addition were explored; for instance, adding all the calcium carbonate to the NBR masterbatch in one blend, and distributing it between the NR and the NBR in another blend. In most cases, the differences in properties was minimal, therefore it was decided to leave the filler in the NBR masterbatch.

Swelling in butter oil could only be reduced to acceptable levels by reducing the quantity of the highly swelling component in the blend, the NR. Thus, 35:65 and 25:75 NR:NBR blends were investigated. A major consequence was that the amount of carbon black in the NR masterbatch could be greatly increased, whilst remaining within the overall limit set for carbon black in the blend. This would assist in lowering the swelling. Given the reduction in NR content of the blend, it was necessary to revert to silica as filler in order to ensure adequate strength properties.

**Table 8.3** Physical properties for 50:50 NR:Breon N41 blends

| *Property* | *Accelerator system (phr)* | | |
|---|---|---|---|
| | *MBS/MBTS (1.24/0.16)* | *MBS/TETD (1.24/0.18)* | *TBTD (1.17)* |
| Tensile strength, MPa | 12.9 | 13.4 | 13.4 |
| Butter oil swelling, % | 36.2 | 42.3 | 34.7 |

These vulcanizates swelled less in butter oil, but the swelling was still excessive, possibly due to the low crosslink density in the NR phase and to insufficient filler. Therefore, the curative levels were doubled and the black level increased. By these means, 25:75 NR:NBR41 blends were formulated to give butter oil swelling well within the specification limit, and tensile strengths were also improved (Table 8.4). Based on this information NR/NBR34 blends were also produced with similar properties (Table 8.4).

A compatibilizer, polychloroprene (CR) was introduced into the formulations for the NR/NBR34 blends; this was based on the work described in Chapter 7. CR (5 phr) replaced a similar quantity of NBR34 and magnesium oxide was added to assist vulcanization of the CR [2–4].

At this stage, a separate CR masterbatch was mixed and included in the cross-blending process. The basis for formulating NR/NBR blends for milking inflations had been established, to the point that only minor changes in the level of the ingredients should be required to meet the specifications. Blends based on NBR containing 34% acrylonitrile were preferred to blends containing NBR with higher acrylonitrile content on both economic grounds and their overall superior properties. The use of calcium carbonate as a filler for the NBR phase did not prove successful; the reduction in strength properties was unacceptable. The eventual selection of a S/CBS/TETD cure system is in accord with the performance of this cure system in gum vulcanizates (Chapter 5). Representative overall formulations emerging from this laboratory investigation and which comply with FDA regulations are given in Table 8.5, and some physical properties of the blends are presented in Table 8.6.

### 8.2.2 Large-scale mixing

The compounds showing most promise, at the time, were mixed on a larger scale at the facilities of a major equipment manufacturer. Both

**Table 8.4** Physical properties for 25:75 NR:NBR blends

| *Property* | *NR/NBR41* | | *NR/NBR34* |
|---|---|---|---|
| | *Accelerator system*[a] *(phr)* | | |
| | *MBS/MBTS (2.48/0.32)* | *MBS/TBTD (2.48/0.46)* | *CBS/TETD (2.6/0.28)* |
| Tensile strength (MPa) | 14.0 | 14.3 | 16.2 |
| Butter oil swelling (%) | 6.2 | 6.3 | 7.7 |

[a] All contain 2.6 phr sulphur

**Table 8.5** Blend formulations meeting US FDA regulations

| *Ingredients* | *35:65 NR:NBR34* | *25:7.5:67.5 NR:CR:NBR34* |
|---|---|---|
| N660 black | 18.5 | 18.4 |
| DOP | 36.5 | 40.5 |
| Silica | 26.7 | 29.3 |
| Magnesium oxide | – | 1.0 |
| Zinc oxide | 5 | 5 |
| Stearic acid | 3 | 3 |
| IPPD | 5 | 5 |
| Wax | 2 | 2 |
| Sulphur | 2.4 | 3.15 |
| CBS | 3.65 | 3.1 |
| TETD | 0.28 | 0.28 |

**Table 8.6** Properties of blends formulated to meet US FDA regulations

| *Properties* | *35:65 NR:NBR34* | *25:7.5:67.5 NR:CR:NBR34* |
|---|---|---|
| Tensile strength, MPa | 12.2 | 9.7 |
| Crescent tear strength, N/mm | 34.2 | 23.8 |
| Hardness, Shore A | 59 | 54 |
| Butter oil swelling, % | 5.5 | 7.3 |

non-compatibilized and compatibilized blends were selected due to the properties being very close to those stipulated by the milking inflation specification. The formulations of NR, NBR and a combined NBR/CR masterbatch are given in Table 8.7, and the overall formulations of the two blends are presented in Table 8.8. The latter are close to the final formulations from the laboratory development programme given in Table 8.5. However, it should be noted that the NBR and CR were mixed in the same masterbatch, thereby reducing the number of mixes, and magnesium oxide was not included in this masterbatch.

The masterbatches were prepared, cross-blended, and then finalized with the curatives in a 47 litre Farrel F50 Banbury mixer fitted with Synchronous Technology (ST) rotors. The two blends were vulcanized at 180 °C to rheometer $t_{max}$. The properties were very good (Table 8.9), in fact better than those of the same blends mixed on a small scale. The unaged properties and butter oil swellings all conformed to the specifications for milking inflations. The improvement in properties seen from

**Table 8.7** Masterbatch formulations for large-scale mixing

| *Mix no.* | 1 | 2 | 3 | 4 |
|---|---|---|---|---|
| SMR L | 100 | – | 100 | – |
| NBR34 | – | 100 | – | 90 |
| CR | – | – | – | 10 |
| N660 | 52.8 | – | 73.5 | – |
| DOP | 28 | 41 | 30 | 41 |
| Silica, VN3 | – | 48 | – | 45 |
| Zinc oxide | 5 | 5 | 5 | 5 |
| Stearic acid | 3 | 3 | 3 | 3 |
| IPPD | 5 | 5 | 5 | 5 |

**Table 8.8** Overall formulations of blends mixed on a large scale

| *Crossblend mix nos.* | *1 and 2* | *3 and 4* |
|---|---|---|
| NR, SMR L | 35 | 25 |
| NBR34 | 65 | 67.5 |
| CR | – | 7.5 |
| N660 | 18.5 | 18.4 |
| DOP | 36.5 | 38.3 |
| Silica, VN3 | 31.2 | 33.8 |
| Zinc oxide | 5 | 5 |
| Stearic acid | 3 | 3 |
| IPPD | 5 | 5 |
| Sulphur | 2.86 | 3.64 |
| CBS | 2.6 | 2.6 |
| TETD | 0.28 | 0.28 |

**Table 8.9** Properties of blends mixed on a large scale

| *Properties* | *35:65 NR:NBR* | *25:7.5:67.5 NR:CR:NBR* |
|---|---|---|
| Hardness, Shore A | 50 | 52 |
| Tensile strength, MPa | 14.4 | 12.9 |
| Elongation at break, % | 595 | 530 |
| Crescent tear strength, N/mm | 30 | 17 |
| Tension set, % | 2 | 1.7 |
| Butter oil swelling, % | 3 | 0.3 |

this larger-scale mixing trial augered well for production of these blends on a full factory scale.

### 8.2.3 Industrial experience

The approaches adopted here, that is, selection of the best of the available cure system on the basis of crosslink distribution and the use of compatibilizer, were included in the development programme of a major manufacturer of milking inflations experiencing difficulty in meeting a specification regardless of elastomer selection. It proved possible to meet the specification with a NR/NBR blend and compound was eventually mixed at a factory scale. Milking inflations were injection-moulded and passed simulated service testing. Full-scale field trials were also successful.

There can be no better test of a technology than this, nor a better confirmation of the viability of these blends when compounded correctly from a position of good knowledge of the problems they present and appropriate solutions.

## 8.3 FORMULATIONS FOR MILKING INFLATIONS TO MEET BGA REGULATIONS

Within the European Union, the dominant legislation affecting milking inflations is the German Food Law. Recommendations issued by the Bundesgesundheitsamt (the Federal Health Authority or BGA) are widely respected, and these incorporate lists of approved ingredients for particular applications. Acceptable concentrations are also defined. The section which affects the compounding of rubbers is Recommendation 21. This divides the applications of rubber in contact with food into five categories, depending upon the degree of contact. The one relevant to milking inflations is Category 3: short-term contact, not more than 10 minutes. The regulations differ considerably from those promulgated by the US FDA, and necessitate the development of almost completely different formulations.

The BGA regulations permit the use of 30% w/w furnace black rather than the 10% w/w allowed by the FDA. Neither polychloroprene rubber nor sulphenamide accelerators meet with BGA approval. However, Heveaplus MG30 is acceptable to the BGA, and may be used as a compatibilizer. In studies of gum vulcanizates, good physical properties were obtained with 5% or 10% of the MG30 incorporated in NR/NBR blends (Chapter 6), with the best properties being obtained when part of the NBR is replaced by the MG30. Changes to the accelerators and plasticizers are also required to meet BGA requirements.

MBTS and DOTG (di-*o*-tolylguanidine) were investigated as accelerators in the vulcanization of blends formulated to meet BGA regulations. The plasticizer DOP was replaced with liquid paraffin, which has a permitted limit of 5% w/w. Following the experience of formulating to FDA regulations, the compounds were based on 35:65 and 25:75 NR:NBR ratios. The use of the higher permitted level of carbon black causes less difficulty in meeting strength requirements, and it is possible to use calcium carbonate as a filler rather than the more expensive silica.

Compounds containing 30–35 phr carbon black loaded in the NR phase, 55 phr calcium carbonate in the NBR phase and the maximum amount of liquid paraffin (11 phr overall) were vulcanized with 2.2 phr sulphur, 2.7 phr MBTS and 0.2 phr DOTG at 180 °C. Whilst strength properties were acceptable (tensile strength > 11 MPa and tear strength > 22 N/mm), swelling in butter oil was about 20%. The latter difficulty arises largely from the low limit on plasticizer. Time did not permit further development to reduce butter oil swelling, but experience in formulating to meet FDA regulations suggests that the necessary combination of low swelling and strength should be attainable.

## 8.4 CONCLUSION

Building on the understanding of requirements in respect of appropriate cure systems and compatibilizers for NR/NBR blends, practical formulations have been developed within the context of milking inflations. The constraints imposed on the range of ingredients which may be used in this food contact application make this a particularly difficult task. Nonetheless, it has been possible to almost meet a demanding specification for a milking inflation whilst complying with the US FDA regulations. Representative formulations have also been mixed at a scale close to factory practice without difficulty, and vulcanizates had better properties than those of comparable compounds mixed at a much smaller laboratory scale.

Furthermore, the developments on which this work is founded have been successfully implemented by a manufacturer of milking inflations. Compounds have been mixed at a factory scale and processed into milking inflations which have successfully passed full field trials.

The extensive changes to formulations necessary when developing materials for markets governed by BGA regulations were addressed. The stringent limit on plasticizer loading causes difficulty in attaining the necessary low swelling in butter oil, and further work will be necessary to overcome this.

## 8.5 REFERENCES

1 United States (1980) *Federal Register Code of Federal Regulations. Part 21, Food and Drugs, Parts 170 to 199*. Washington, DC: US Government Printing Office.
2 Franta, I. (1989) *Elastomers and Rubber Compounding Materials*, Elsevier, Amsterdam.
3 Brydson, J.A. (1988) *Rubbery Materials and their Compound*, Elsevier, London.
4 Murray, R.M. and Thompson, D.C. (1964) *The Neoprenes* (International Edition), E.L. du Pont de Nemours, Wilmington.

# 9

# Novel natural rubber/ethylene propylene copolymer (EPM) blends

*Paul S. Brown*
*Tun Abdul Razak Research Centre, MRPRA, UK*

## 9.1 INTRODUCTION

Blends of NR with ethylene propylene copolymer (EPM) which have the excellent ozone resistance of EPM, yet contain a significant amount of NR, should enjoy an economic advantage over current EPM formulations as EPM are relatively expensive elastomers. There may also be some property benefits. The main applications for such materials would be in the cable industry as electrical insulation sheathing. The economic benefits should, of course, be greatest in the natural rubber producing countries of the world where the partial replacement of an imported synthetic polymer with a locally sourced material gives additional financial benefits.

The formulation for a typical high tension cable sheathing compound is given in Table 9.1. For use by the cable insulation industry the materials should be capable of being crosslinked by high energy radiation, such as electron beam irradiation. Other potential markets for NR/EPM blends with good physical properties should be in the automotive and construction industries as seals.

For a blend of NR and EPM to show the excellent ozone resistance of the EPM component probably means that the EPM should form a continuous phase while the NR component may form a disperse phase, or

*Blends of Natural Rubber*. Edited by Andrew J. Tinker and Kevin P. Jones.
Published in 1998 by Chapman & Hall, London. ISBN 0 412 819406

**Table 9.1** Typical high tension cable sheathing compound

| | |
|---|---|
| EPM | 100 |
| LDPE | 10 |
| Zinc oxide | 5.0 |
| Red lead | 5.0 |
| Calcined clay | 45 |
| Paraffinic oil | 5 |
| Paraffinic wax | 5 |
| Antidegradant | 1.5 |
| Silane A 172[a] | 1 |
| Liquid 1,2-BR | 6.0 |
| Peroxide (40%) | 6.0 |

[a] Osi Specialities, Switzerland

may itself be continuous. Gum mixes of these two polymers with the polymer ratio varying from 1:3 through to 3:1 (by weight) were found to possess a disperse phase/matrix phase blend morphology. Figure 9.1 shows a phase contrast light micrograph of a conventional 50:50 NR:EPM blend at approximately 400 times magnification. Under these imaging conditions NR appears darker than EPM and it is evident that NR forms the matrix phase. The lighter EPM is present as a disperse phase within this matrix. The EPM domains are elongated, roughly 20 × 5 μm in size, and separated by relatively narrow (2–10 μm) regions of NR. The interphase boundary appears to be smooth. Therefore, the blends do not naturally possess the desired morphology. A successful NR/EPM blend requires the modification of this natural morphology.

## 9.2 DYNAMIC VULCANIZATION

Dynamic vulcanization has been used to improve the properties of thermoplastic elastomer blends [1] and to control the dispersion and distribution of carbon black in polymer blends [2]. Dynamic vulcanization is a procedure whereby one of the components of the blend is vulcanized, or partly vulcanized, during the mixing process, whilst the other component remains completely uncured. The introduction of crosslinks into one of the phases increases the viscosity of this phase, leading to a change in blend rheology, and hence in the morphology of the blend [3, 4]. The uncrosslinked phase is unaltered by the vulcanization in the other phase, permitting easy processing of the blend. The modified blend can then be additionally compounded (if required), formed to shape and a second cure system used to 'fix' the finished article.

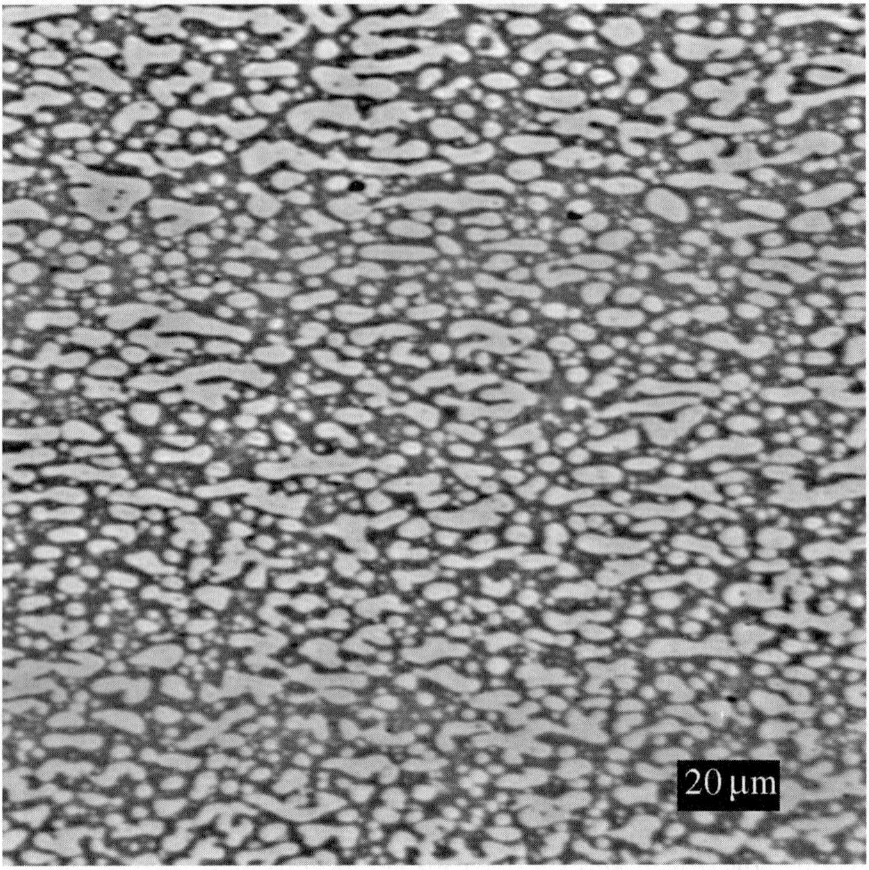


**Figure 9.1** Phase contrast micrograph of a conventional NR/EPR blend.

Because dynamic vulcanization takes place inside the mixer, there is an economic requirement for the mix cycle to be as short as possible. Therefore, there are several constraints forced upon the dynamic cure system, which must:

- be specific to one phase of the blend
- be very active and fast with little delay
- not interfere with a second stage cure
- permit subsequent processing of the blend
- be based on standard rubber industry chemicals

This last point is important since it allows the technology to be readily available.

The observation that the blend morphology comprises a NR matrix means that the NR phase has the lower viscosity under the mixing conditions [3]. It is, therefore, necessary to increase the NR phase viscosity, or reduce the EPM phase viscosity, to effect the morphology change required. Achieving cure system specificity in favour of the NR component of a NR/EPM blend is relatively easy. EPM, being a copolymer of two olefin monomers (ethylene and propylene), is a fully saturated elastomer: thus it is not crosslinked by the accelerated-sulphur cure systems commonly employed in the vulcanization of NR. The subsequent vulcanization of the blend using a radical crosslinking system should occur in both phases, although residues from the sulphur cure may act to reduce the efficiency of this process. Dicumyl peroxide (Dicup) was used as the main radical crosslinking agent in the laboratory, although electron beam irradiation was also performed.

The feasibility of using dynamic vulcanization in this way had been demonstrated in earlier work [5]. A system was developed that was both selective and fast, but the subsequent peroxide vulcanization of these blends was poor. The first priority was thus to develop an accelerated-sulphur cure system for the NR phase of the blend, which did not unduly affect the subsequent radical induced crosslinking of the blend.

## 9.3 SELECTION OF DYNAMIC CURE SYSTEM

A wide range of accelerated sulphur cure systems was studied, some 50 in total. The accelerator:sulphur ratios were chosen to cover the whole spectrum of cure system efficiency from high sulphur:low accelerator conventional cure systems through to low sulphur:high accelerator efficient cure systems. Single polymer NR compounds were produced by mill finalizing a masterbatch of NR, zinc oxide and stearic acid with the appropriate curatives. The compounds were vulcanized to optimum cure as determined by cure-rheometry at 150 °C using Monsanto ODR and MDRE rheometers. The data from the rheometers was also used to assess the rate of vulcanization and the initiation period. Table 9.2 gives details of some of the curative systems studied.

The efficacy of a second-stage peroxide vulcanization was assessed using unextracted samples of the cured sheets to model the effect of the curative residues on the subsequent radical cure. A weighed sample of the NR vulcanizate (c. 25 × 25 × 1 mm) was painted with a liquid peroxide (2,5-dimethylhexane-2,5-di-*t*-butyl peroxide, a peroxide with a similar decomposition half-life to that of Dicup). After leaving the sample for 16 hours to allow for the peroxide to become homogeneously distributed throughout the sample, it was reweighed to determine the exact peroxide loading. The doped samples were returned to the press for an additional vulcanization period of one hour at 160 °C.

**Table 9.2** Dynamic curing systems

| *Cure system* | *Cure time 150 °C (min)* | *Peroxide secondary (cure)* |
|---|---|---|
| 0.75 S, 0.75 CBS, 0.1 ZDEC | 11.5 | Poor |
| 0.3 S, 0.1 TMTM, 0.1 Robac $\gamma$[a] | 10 | Poor |
| 0.75 S, 0.35 CBS, 0.35 ZDMC | 4 | Good |
| 0.75 S, 0.35 CBS, 0.35 ZBED | 9 | Good |
| 0.5 S, 0.3 TMTD | 7.2 | Good |
| 1.0 S, 0.4 CBS, 0.2 TMTD | 9.4 | Good |
| 0.75 S, 0.75 ZDMC, 0.2 CBS | 6.3 | Good |
| 0.75 S, 0.75 ZDMC, 0.2 DPG | 2.5 (5.8)[b] | Good |
| 0.75 S, 0.75 ZDMC, 0.2 Robac ZIX[a] | 8 | Good |
| 0.75 S, 0.75 ZDMC, 0.1 DPG | 2.3 (6.0)[b] | Good |
| 0.75 S, 0.75 ZDMC, 0.3 DPG | 2.0 (5.7)[b] | Good |

[a] Robinson Brothers, UK
[b] Brackets indicate cure times at 140 °C

The crosslink densities of all of the samples, both original and doped, were interpolated from volume swelling measurements in chloroform using a previously determined crosslink density:swelling ratio correlation. The efficacy of the secondary peroxide cure was then estimated by a comparison of the observed increase in crosslink density in a particular NR vulcanizate with that of a similarly treated Dicup-cured NR control sample. The systems were ranked as good, fair or poor (Table 9.2). Three very fast systems based on sulphur, ZDMC and DPG, all with cure times below three minutes at 150 °C, and a fourth, slightly slower system using sulphur, ZDMC and CBS were developed using this approach (Table 9.2).

## 9.4 DYNAMIC VULCANIZATION OF NR/EPM BLENDS

### 9.4.1 Gum blends

Dynamic vulcanization of the NR phase of a NR/EPM blend was first attempted in the gum state in a Brabender PL2000 mixer with a 350S mixing head and Banbury style rotors (Table 9.3). The only ingredients in the mix were the two polymers, zinc oxide, stearic acid and the cure system. The insertion of crosslinks into the NR phase could be monitored by its effect on the mix viscosity and hence the mixing torque. In the absence of the dynamic cure system there is a gradual fall in mixing torque with mixing time. When curatives were present this torque decay

**Table 9.3** Mixer type and mix batch size

| *Mixer* | *Capacity at 0.80 fill factor (kg)* |
|---|---|
| Brabender PL2000, 350S head | 0.38 |
| BR Banbury | 1.53 |
| 00C Banbury | 4.20 |
| K2A Intermix | 43.8 |
| F50 Farrel | 45.8 |

was halted at the onset of the vulcanization, crosslinking resulting in a more viscous mix. Dynamic vulcanization began once the mix reached a temperature of 125–135 °C, then the mixing torque rose as dynamic vulcanization proceeded (Figure 9.2). The rise in mixing torque during dynamic vulcanization results in an increase in the power input to the mix, thus accelerating the temperature rise. It is this factor (mix temperature) that dictates events in the rest of the mix cycle. The mix has to be dumped below 165 °C to avoid polymer degradation. A change in the tone of the noise of the mixer motor and an increase in ram movement on the mix also accompany dynamic vulcanization.

The morphology of these dynamically vulcanized blends was studied under identical conditions to those used for the conventional blend. The micrograph (Figure 9.3) shows a finer structure than found earlier. In addition, the phases appear to be co-continuous and the phase boundaries appear to be much rougher than in the conventional blend.

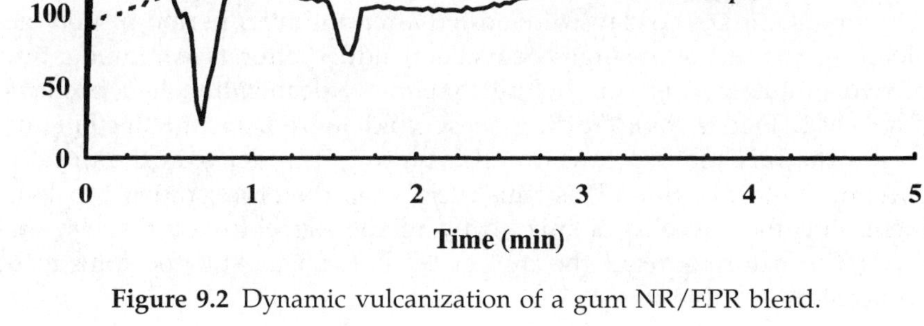


**Figure 9.2** Dynamic vulcanization of a gum NR/EPR blend.

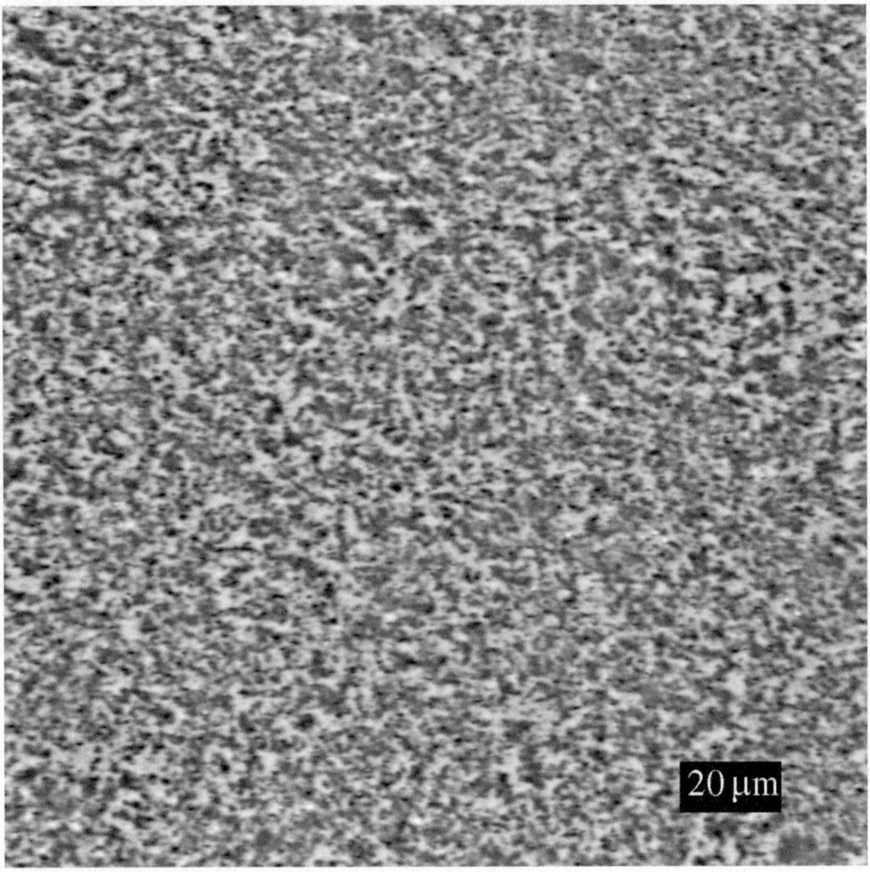


**Figure 9.3** Phase contrast micrograph of a dynamically vulcanized NR/EPR blend.

### 9.4.2 Clay-filled blends

Dynamic vulcanization of clay-filled NR/EPM blends was first attempted using a relatively conventional mix cycle. The powders, curatives and some of the clay were added after an initial period of polymer blending, the rest of the ingredients being added after a total mixing time of two minutes. This was before dynamic vulcanization had occurred (Table 9.4, Figure 9.4). Adding these cold ingredients to the hot mix causes the mix temperature to fall, resulting in a delay in the onset of dynamic vulcanization. This mix cycle was, therefore, much too long. Increasing the severity of mixing by increasing both rotor speed and mixer fill factor reduced the mix cycle, but it was still too long to be economically viable.

**Table 9.4** Mixing cycle

| *Time* | *Action* |
|---|---|
| Start | Add polymers |
| 30 s | Add powders and curatives and some clay |
| 2 min or at DV cure | Add clay and oil |
| 165 °C max. or after 1 min mixing | |
| After the last ram up period | Dump |

The optimum time to perform the addition of the second portion of clay was found to be just after the torque rise due to dynamic vulcanization, rather than before it. This new mix cycle avoids the delay in the onset of dynamic vulcanization and saves about two minutes' mixing time (compare Figures 9.4 and 9.5: the latter mix could have been dumped a minute earlier than shown). Adding some of the clay and all of the oil after dynamic vulcanization is beneficial to mixing. The addition of the cold ingredients lowers the mix temperature by about 20 °C, whilst the presence of oil lowers the mix viscosity, hence reducing mixing torque and the rate at which the temperature subsequently increases. Mixing is thus able to continue for a sufficient time after the second addition to ensure that the materials are properly dispersed. The total mix cycle is about six minutes in length.

A small reduction in mixing time was achieved when scaling up to a 00C Banbury mixer, but a sweep stage was required (Figure 9.6). The

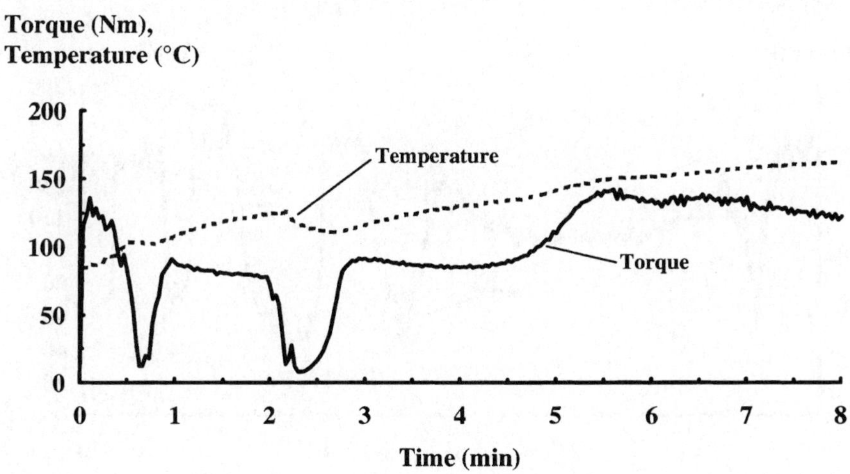


**Figure 9.4** PL2000 Brabender mixing record for clay-filled mix; oil added after two minutes.

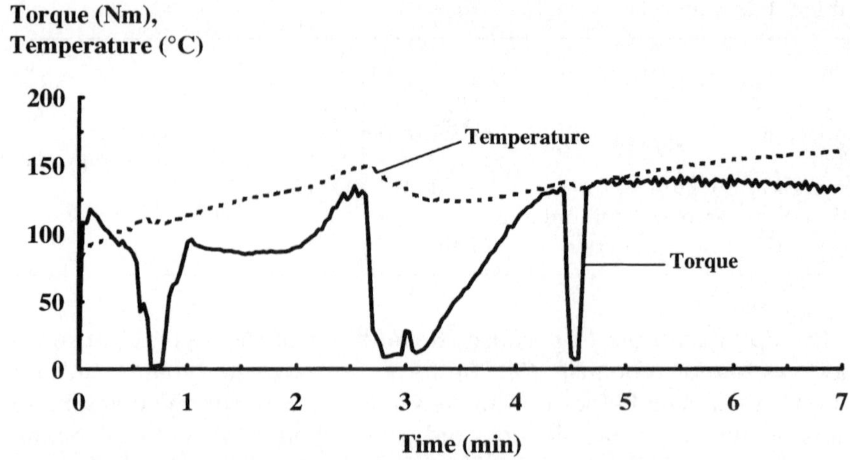


**Figure 9.5** PL2000 Brabender mixing record for a clay-filled mix; oil added after the torque maximum.

total mix cycle was reduced to 4.5–5 minutes in length by an increase in both the rotor speed and the circulating water temperature, and by increasing the amount of clay in the first addition. These changes brought forward the onset of dynamic vulcanization from the 2.5 minutes shown in Figure 9.6 by roughly 30 seconds and also removed the need to sweep the ram, saving a further 30 seconds.

The dynamically vulcanized (DV) mixes were analysed to determine the gel content of each mix by swelling in both THF and chloroform. The

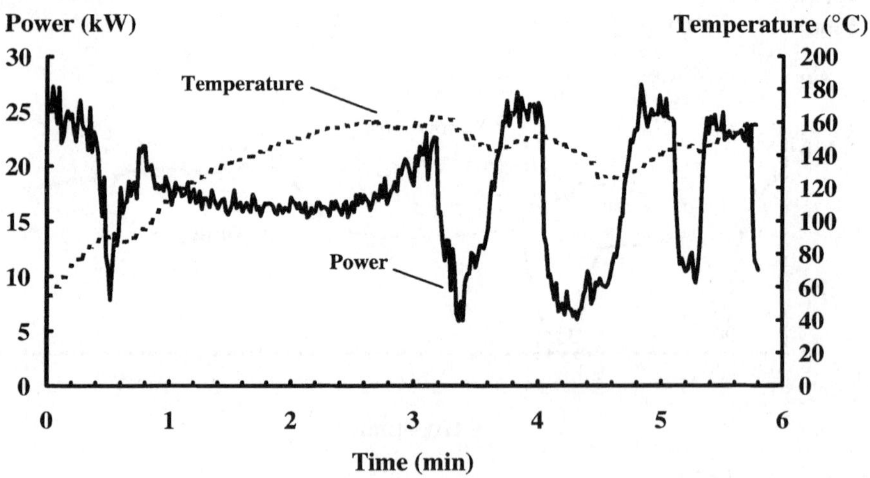


**Figure 9.6** Dynamic vulcanization in a 00C Banbury.

gel content was found to be broadly equivalent to the NR content of the mix, suggesting most of the NR is gelled and most of the EPM is dissolved out of the mixes. This high gel content suggested that dynamic vulcanization was producing an excessive crosslink density in the NR. Further evidence to support this view was provided by the behaviour of the DV blends during processing: band formation on the mill was poor. A portion of each of these DV mixes was finalized with Dicup and cured. These vulcanizates were analysed using swollen-state NMR spectroscopy [6]. After making due allowance for the crosslinks introduced by the peroxide cure, the crosslink density in the NR phase of these blends was rather high, suggesting that the level of the dynamic curatives was too high. A reduction in the DV curative level improved both the processing of the dynamically vulcanized blends and the physical properties of the peroxide-cured vulcanizates. The torque rise on dynamic vulcanization was reduced by this change and became less pronounced.

## 9.5 THE SANDWICH MIX CYCLE

Most of the development mixing was performed in a 380 g internal mixer (Table 9.4). Scale-up to larger sized laboratory mixers was relatively uneventful. A major problem was encountered with the 45 litre mixer. The maximum rotor speed was found to be too low to rapidly generate the temperature rise necessary to initiate dynamic vulcanization and mixing took over eight minutes. The loss of effective mixing during the time when the ram is raised to allow material to be added to the mix also contributed to the length of the cycle, the ram movement being appreciably slower than on the smaller mixers.

The sandwich mix cycle (Table 9.5) was developed to reduce the number of these 'ram up' periods during the cycle. Sandwich mixing is based upon the upside-down mixing cycle where all of the ingredients

**Table 9.5** Sandwich mixing cycle

| *Time* | *Action* | *Fill factor* |
|---|---|---|
| Start | Add polymers, powders, curatives and half clay | 0.60 |
| After DV cure | Add the oil, liquid BR, co-agents, coupling agents and the rest of the clay | 0.80 |
| 3–5 min | Dump | |

are put into the mixer at the beginning of the cycle, powders first. As mentioned above, the addition of clay and oil after dynamic vulcanization has occurred is beneficial to mixing; therefore, it was necessary to retain this step. A cycle was developed in which half of the clay and all of the plasticizers are added after the dynamic cure. The ingredients are thus added in the following order:

1. clay and powders
2. polymers
3. clay and oil

hence the description 'sandwich'. This mix cycle was found to offer considerable time savings when used in the smaller mixers (Figure 9.7), the mix cycle being reduced to only 3–4 minutes.

Keeping back a considerable portion of the mix until after dynamic vulcanization has one negative aspect; the fill factor during the initial period of mixing is considerably less than it is at the end of the cycle. Thus a compromise has to be reached between the conflicting requirements of a rapid temperature rise to initiate dynamic vulcanization and the inferior mixing that can result from an overfull mixer. For the mixers employed during the study a fill factor of 0.8 (times the chamber volume) for the total mix was found to be optimum; this compares with the more usual 65–73% fill factor. Even using this high final fill factor results in the mixer being only 60% full during the period before dynamic vulcanization occurs, the period during which the temperature is required to reach vulcanization temperatures (Table 9.5). This optimum factor will obviously vary with mixer and rotor design.

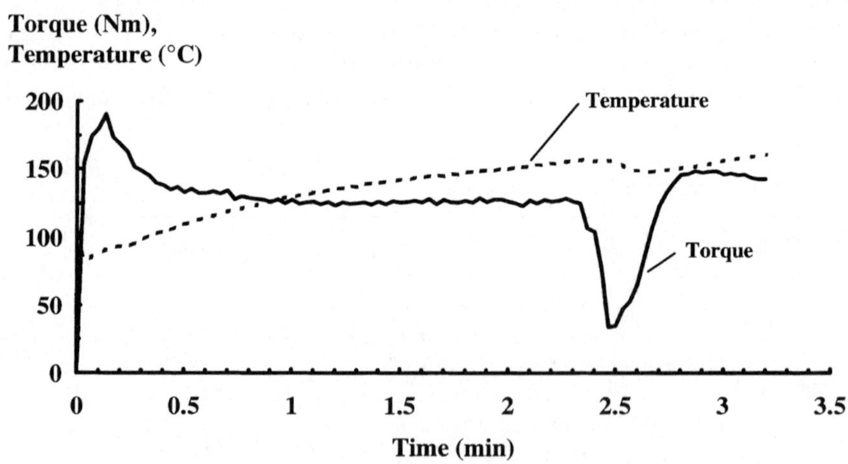


**Figure 9.7** Sandwich mix cycle in a PL2000 Brabender.

Scale-up to pilot plant scale was also attempted using a variable speed Farrel F50 Banbury mixer with ST (Synchronous Technology) rotors. There were some problems with controlling the non-standard oil injection system of this particular mixer. The oil injection system was set up for oils of a much higher viscosity than the 2280 grade used in these mixes, resulting in the injection of the oil at the wrong time in the cycle, and at too high a rate. It had been intended to add the oil at the same rate as it was incorporated, thus maintaining a high mixing torque throughout the cycle. This was not possible, the resultant cycles were unacceptably long, taking up to 7.5 minutes. To facilitate a short cycle the oil loadings were reduced with some success. Optimum mixing conditions were not achieved in this pilot scale mixing.

## 9.6 PROCESSING BEHAVIOUR

The study of optimization of the DV blend processing behaviour used the basic formulations of Table 9.6. Two machines were employed; a 30 mm extruder with two temperature controlled zones (the barrel and the head) and a 60 mm extruder with four temperature controlled zones (the screw, two barrel zones and the head). Both extruders can have the feed area heated to the same temperature as the neighbouring barrel zone, but this was not found to be beneficial. The 60 mm extruder has a choice of three screw configurations: conventional helical, barrier and pin screws. The optimum temperatures for the various temperature controlled zones are given in Table 9.7. The pin screw gave the best results in

**Table 9.6** Basic formulation

| *Ingredient* | *Loading (phr)* |
|---|---|
| SMR CV | 40–60 |
| EPM | 60–40 |
| Zinc oxide | 5 |
| Stearic acid | 2 |
| Sulphur | 0.13 |
| ZDMC | 0.13 |
| DPG | 0.02 |
| Titanium dioxide (Rutile) | 10–15 |
| Clay | 60–100 |
| Oil | 20–30 |
| Silane | 0–1.0 |
| Liquid 1,2-BR | 5 |
| Co-agent | 0–1.0 |

**Table 9.7** Extrusion conditions: zone temperatures

| *Zone* | *30 mm* | *60 mm* |
|---|---|---|
| Die head | 85–95 °C | 85–95 °C |
| Barrel 1 | 75–80 °C | 80–90 °C |
| Barrel 2 | | 70–80 °C |
| Screw | | 65–75 °C |
| Feed | Unheated | Unheated |

the 60 mm extruder, the optimum pin configuration used all six pins in each of the seven (of ten) rows nearest to the die head.

The polymer ratio was varied from 4:6 to 6:4 (NR:EPM) using three EPM of different ethylene content (Table 9.8). The higher the ethylene content of the EPM, the faster the temperature rises during mixing (EPM viscosity increases with ethylene content). The DV blends containing Vistalon PE805 (the highest ethylene content EPM) gave poor extrudates using the Garvey die of ASTM D2230; both the surface finish and the sharp-edge definition of the extrudate scored only 2 or 2.5 (out of 4). These values were worse than in the DV blends containing either of the other grades of EPM, which both gave good extrudates. Changing the polymer ratio (within this narrow range) appeared to have little effect on extrusion performance.

The type of clay used in the blend (china clay, calcined clay or silane pre-treated calcined clay) has little effect on the mixing process or on the extrusion of the DV blends. Altering the clay level also has little effect on the processing behaviour of the DV blends providing there is a concomitant change in oil level. The extrudate swell is slightly reduced when clay and oil are increased together, improving the Garvey score by a half of one point. If the oil level is not increased with clay load, or if the oil loading is reduced, then extrusion performance is worsened by up to three points, with swelling and surface finish being particularly affected.

**Table 9.8** Ethylene content of the EPM

| *EPM grade* | *Ethylene content* | *Viscosity $M_L(1+4)$ 100 °C* |
|---|---|---|
| Dutral Co 034[a] | 70% | 40 |
| Dutral Co 054[a] | 55% | 40 |
| Vistolan PE 805[b] | 75% | 35[c] |

[a] Enichem Elastomeri, Italy
[b] Exxon
[c] Measured at 125 °C

**Table 9.9** Garvey die ratings

| *Property* | *Score* |
|---|---|
| Swelling | 2–3 |
| Edge | Usually 3–4 |
| Smoothness | Usually 4 |
| Corners | Usually 4 |
| Total | 13 or better |

Increased oil loading extends the mixing time required for oil absorption and thus the total mix cycle. Variation in the loadings of the other ingredients (silane, co-agent or titanium dioxide) had very little effect on either mixing or extrusion.

The Garvey die ratings are summarized in Table 9.9. The highest score attained was 15 for an unvulcanized control blend; its dynamically vulcanized analogue scored 13. The best rating for a DV blend was 14.5, a 60:40 (NR:EPM) blend with 60 phr clay and 20 phr oil. DV blends mixed in the F50 scored up to 14 despite their slightly lower oil levels (16 phr). The typical HT cable insulation compound of Table 9.2 scores 11.5 [7], with extrudate swell being a particular weakness. It would thus appear that these materials possess acceptable processing behaviour for the target market.

## 9.7 VULCANIZATION OF DV BLENDS

### 9.7.1 Peroxide

The DV blends have adequate processing behaviour for shaping the finished product, although they do need to be vulcanized to maintain that form. As noted above, EPM vulcanization requires the use of a radical based cure system. The sulphur-cure system used for the dynamic vulcanization of the NR component was selected both for its speed of action and for its minimal effect on a subsequent radical cure. It is, however, a common industrial practice to employ co-agents to increase the yield of crosslinks during the vulcanization of EPM, either with radiation or peroxide cures. This improves the physical properties of the finished product. A series of such co-agents (Table 9.10) was investigated, both singly and in pairs, to try to optimize Dicup curing of the DV blends. The peroxide was used at 2.5 phr (Table 9.11) and the sheets were vulcanized for one hour at 160 °C (c. six half-lives). The hoped for synergism between pairs of co-agents was not observed. No particular co-agent or pair of

**Table 9.10** Radical co-agents

| *Co-agent* | *Comments* |
|---|---|
| Silane RC 1[a] | Allyl silanes, couple to the clay and act as co-agents |
| Silane A 172[a] | |
| Liquid BR (High vinyl) | Plasticizes the mix too |
| HVA-2 (*m*-phenylamine bismaleimide)[b] | Gives high modulus, low TS |
| SR 350 (Trimethylolpropane trimethacrylate, TMPTMA)[c] | Resin monomer/co-agent |
| TAC (tri-allylcyanurate)[d] | Hindered the electron beam cure |

[a] Osi Specialities, Switzerland
[b] DuPont
[c] Sartomer, USA
[d] Rhein Chemie, Germany

**Table 9.11** Tensile data

| *Co-agent* | *T.S. (MPa)* | *E.B. (%)* | *MR 100 (MPa)* | *Comp. set. 1d/100 °C (%)* |
|---|---|---|---|---|
| | | 40:60 NR:EPM | | |
| None | 10.4 | 626 | 1.25 | 22 |
| HVA-2 | 6.92 | 494 | 1.34 | 24 |
| HVA-2, RC 1 | 6.68 | 502 | 1.33 | 22 |
| HVA-2, SR 350 | 6.49 | 483 | 1.31 | 21 |
| RC 1, SR 350 | 11.2 | 622 | 1.27 | 23 |
| RC 1 | 11.3 | 635 | 1.23 | 24 |
| SR 350 | 11.9 | 626 | 1.21 | 23 |
| | | 50:50 NR:EPM | | |
| HVA-2 | 6.55 | 451 | 1.38 | 20 |
| HVA-2, RC 1 | 7.61 | 507 | 1.35 | 20 |
| HVA-2, SR 350 | 7.48 | 492 | 1.31 | 20 |
| RC 1, SR 350 | 11.2 | 600 | 1.27 | 21 |
| RC 1 | 10.6 | 599 | 1.27 | 21 |
| SR 350 | 10.8 | 575 | 1.26 | 21 |

Co-agents present at 1 phr total, at a 50:50 ratio where mixed co-agents are used. Vulcanized using 2.5 phr Dicup. Compounds from the basic formulation of Table 9.6 using Dutral Co 034, 60 phr Polestar 300 R clay and 20 phr Strukpar 2280 oil co-agents showed significantly superior physical properties, although the presence of HVA-2 gave an increased modulus, but at the expense of a significant reduction in tensile strength (Table 9.11). Of the two allyl silanes investigated, Silquest RC 1 is to be preferred for reasons of health and safety: the by-product of its coupling reaction with clay is ethanol whilst that of Silquest A 172 is a more toxic alcohol.

### 9.7.2 Radiation

Electron beam irradiation was performed at the Nuclear Energy Research Unit in Malaysia and also at a commercial facility in the UK. The radiation dose was varied from 100 to 500 kGy (1 kGy = 1 $JM^{-3}$). Dosages used in Malaysia were too low to give adequate crosslinking, but served to establish conditions for the later work. The beam current used at the UK facility was very high, 9 kW focused to a 2 cm diameter disc. This high-energy density beam produced excessive heating in the sample sheets, causing porosity and loss in ultimate properties. Based on the observed modulus, the optimum dosage was found to be about 300 kGy, but this may have been influenced by the degree of porosity. Comparing moduli of irradiated and Dicup cured sheet suggests that electron beam irradiation is giving considerable crosslinking. Despite its use as a co-agent in electron beam irradiation of EPM compounds, tri-allyl cyanurate (TAC) was found to reduce the crosslinking efficiency when used in these DV blends; Silquest RC 1 and the Sartomer resin are the preferred agents. Thus irradiation is a feasible method of vulcanization of the DV blends, although the process has not been optimized.

## 9.8 RESISTANCE TO ENVIRONMENTAL DAMAGE

The main aim was to produce NR/EPM blends with the ozone resistance of EPM. Ozone is not, however, the only aggressive agent in the environment. Polymers have to be resistant to attack by oxygen, operate at temperatures above ambient and may suffer attack by the UV component of natural daylight.

### 9.8.1 Hot air oven ageing

The suitability of a blend for operation at elevated temperatures is assessed by hot air oven ageing. An industrial specification for a range of marine cables requires that the change in tensile properties is less than 30% after ageing for 7 days at a temperature 30–40 °C above the operating temperature. All of the DV vulcanizates tested passed this specification for ageing at 100 °C, but all failed at 125 °C. Thus the upper limit for cables produced from these materials is 60–70 °C.

The formulations tested, unlike the typical EPM compound of Table 9.1, contain no antioxidants. Correct compounding should improve the heat ageing of the DV blend vulcanizates and thus raise their operating temperature. The antioxidants could be added at any time in the cycle, although after the DV cure might be the best option.

### 9.8.2 Ozone resistance

Ozone resistance testing was performed to ISO 1431/1: (1989) A, strained to 20% under an atmosphere containing 50 ppm ozone for 72 hours at 40 °C in the dark. The polymer ratio was varied from 6:4 to 4:6 (NR:EPM). All of the EPM rich DV blend vulcanizates showed no cracking during the test period (40:60 and 45:55 polymer ratio). The 60:40 NR rich blends all showed cracking; in most cases patches of fine cracks were first observed after 48 hours, the cracks growing during the remaining period of exposure. Only a few 55:45 polymer ratio blends were produced and these showed no cracking. The 50:50 blends generally passed the test. However, some of the blends mixed at the pilot plant scale did develop cracks during the last 24 hours of exposure. These were the blends that had the most problems during mixing so it is probable that the blend morphology was not the required one, and that correct mixing would give no failures. These results are summarized in Table 9.12.

### 9.8.3 UV light ageing

The resistance to daylight ageing was examined using high intensity, daylight spectrum light produced from a xenon tube using appropriate infra red and UV filters in a Heraeus Xenotest 150S environmental chamber. The test samples were rotated 180° away from the light every

**Table 9.12** Summary of ozone resistance

| *Polymer ratio (NR:EPM)* | *Result* |
|---|---|
| 60:40 | All fail, cracks appear within 48 hrs |
| 55:45 | All pass, no cracks observed |
| 50:50 | Most samples pass, but some of the samples mixed on a large scale developed cracks during the last 24 hrs of exposure |
| 45:55 | All pass, no cracks observed |
| 40:60 | All pass, no cracks observed |

**Table 9.13** Formulations subjected to ageing under light

| *Ingredient* | *DV blend* | *EPM control* |
|---|---|---|
| SMR CV | 40 | 0 |
| Dutral Co 034 | 60 | 100 |
| Clay | 40 | 40 |
| Strukpar 2280 oil | 20 | 20 |
| Titanium dioxide (Rutile) | 20 | 20 |
| Zinc oxide | 5.00 | 5.00 |
| Stearic acid | 2.00 | 2.00 |
| Sulphur | 0.13 | 0 |
| ZDMC | 0.13 | 0 |
| DPG | 0.02 | 0 |
| TAC | 1.50 | 2.00 |
| Dicup | 1.00 | 2.00 |

Sandwich mix cycle in 00C Banbury. Dicup added on mill, sheet cured 160 °C one hour

10 s in order to reduce sample heating. Although the instrument used has three light intensity settings, sample heating was found to be too great at all but the lowest value. At this light intensity the samples were generally only 5–10 °C above ambient temperature, whereas at the maximum light setting sample temperatures were as high as 70 °C (40–50 °C above ambient). This low intensity setting still provides relatively high intensity light, being about 15 times the average light intensity at the earth's surface [8], allowing for cloud cover and haze. The long duration of the experiment (2000 hrs) requires that the samples were masked to provide two control samples of the same thermal history, even at the lowest light intensity chosen for the testing.

A typical DV blend was compared with a similar EPM vulcanizate (Table 9.13). Samples were aged for periods of 250–1000 hrs, spending twice the period in the chamber due to the sample rotation. All samples showed a slight discolouration of the exposed surface together with the formation of a skin. The EPM compound had a smooth, shiny skin, whereas that formed on the DV blend was smooth but dull. Whilst the modulus and tensile strength of the DV blend fell, so did those of the control sample, hence the effect of the light ageing was minimal. The properties of the EPM compound were more stable (Table 9.14).

## 9.9 PREFERRED FORMULATION

The preferred formulation is given in Table 9.15 and the properties of this formulation when cured using 2.5 phr Dicup are given in Table 9.16. This

**Table 9.14** Light ageing data, 1000 hrs exposure

| | DV blend | | EPM control | |
|---|---|---|---|---|
| | *Exposed* | *Control* | *Exposed* | *Control* |
| M300, MPa | 2.03 | 2.02 | 2.44 | 2.41 |
| % retention vs original | 87 | 87 | 103 | 101 |
| % retention vs control | 100 | | 101 | |
| Tensile strength, MPa | 4.06 | 4.13 | 12.7 | 10.1 |
| % retention vs original | 75 | 77 | 105 | 84 |
| % retention vs control | 98 | | 126 | |
| Extension at break, % | 809 | 808 | 900 | 881 |
| % retention vs original | 108 | 108 | 97 | 95 |
| % retention vs control | 100 | | 102 | |

**Table 9.15** Preferred formulation[a]

| *Ingredient* | *Loading (phr)* |
|---|---|
| SMR CV | 50 |
| Dutral Co 034[b] | 50 |
| Zinc oxide | 5 |
| Stearic acid | 2 |
| Sulphur | 0.13 |
| ZDMC | 0.13 |
| DPG | 0.02 |
| Titanium dioxide (Rutile) | 10 |
| Polstar 300R clay[b] | 60 |
| Strukpar 2280 oil[b] | 20 |
| Lithene AH[b] | 5 |
| Silquest RC 1[b] | 0.5 |
| SR 350[b] | 0.5 |

[a] It is recommended that antidegradants are added to this formulation for practical applications, see text. For black-filled compound omit the titanium dioxide and replace with 10 phr carbon black.

[b] Proprietary names are given in the table as these were the actual materials used during this work. Their use must not be taken as a recommendation for particular materials; substitution by equivalent materials should have no effect on mixing and properties.

formulation contains no antidegradant; the inclusion of one would be expected to improve heat ageing, possibly sufficiently to extent the operating temperature range. The calcined clay and Silquest RC 1 could be

**Table 9.16** Properties of preferred blend

| | *Original value* | *Aged 7 d at 100 °C* | *% retention* |
|---|---|---|---|
| Tensile strength, MPa | 11.2 | 10.2 | 91 |
| Extension at break, % | 600 | 604 | 101 |
| M300, MPa | 2.88 | 2.88 | 100 |
| Compression. set, 1d/100 °C, % | 21 | 24 | 114 |

replaced with a silane pre-treated clay such as Polarite 103 A without a detrimental effect on properties. This formulation meets all of the specifications for the marine cables, as set out by one UK manufacturer of such material.

Carbon black filled DV blends (15 phr N660 black instead of the titanium dioxide) were also studied. There appeared to be no major difference in the mixing and processing of such blends when compared to their white-filled analogues. Black-filled DV blends, to a formulation analogous to that of Table 9.15, would be expected to show similar behaviour. The use of coloured pigments has not been attempted, but no problems are anticipated.

### 9.9.1 Effect of modifications to the formulation

Calcined clay is to be preferred for the lower compression set it produces and for the reduced water absorption of such clays. Silane pre-treated clays can be used providing that the silane is compatible with a radical cure, i.e. an allyl silane. The level of silane coupling agent added to the mix can be reduced to compensate for the agent used in the pre-treatment process.

The level of clay and oil in the formulation influence both the ultimate properties of vulcanizates and the mixing of the DV blends. Increasing both clay and oil together reduces the ultimate tensile properties, lengthens the mix cycle due to a prolonged oil incorporation time, but improves extrusion, particularly with regard to surface finish and extrudate swell. Increasing the clay alone has a lesser effect on the mix cycle, but produces a drier mix with inferior extrusion performance and tensile properties. Increasing the oil alone prolongs the mixing, gives better extrusion but reduces tensile properties. Reducing either ingredient alone has the opposite effect on properties to that given above.

EPM needs to be present at 50 phr to give the required ozone resistance. The physical and ageing properties of blends at 50:50 and 40:60 polymer ratios (NR:EPM) are similar. Changing the EPM grade has some effect on mixing and properties; this seems to be an effect of the ethylene content of the EPM, although differences in the polymer molecular weight may also have an effect. Low ethylene EPM (55%) gave inferior physical properties, whilst a high molecular weight, high ethylene (75%) was detrimental to the extrusion performance.

## 9.10 CONCLUSION

A natural rubber/ethylene propylene copolymer blend with the ozone resistance of EPM would be potentially suitable for use in the cable insulation industry. In a traditionally prepared NR/EPM blend, the EPM is usually present as a fine dispersed phase in a NR matrix, a morphology considered unlikely to afford the necessary environmental resistance properties. Vulcanization of the NR phase during mixing, dynamic vulcanization (DV), offers a route to altering the blend rheology and thus the blend morphology. For dynamic vulcanization to be successful two things are required – a fast, polymer specific cure system and a rapid temperature rise during mixing. Dynamic vulcanization of the NR component occurs before all of the compound ingredients are added to the mixer, thus higher than usual fill factors are required to achieve the cure temperature in a short mix cycle. Scale-up from large laboratory scale mixers to pilot plant scale was problematic. In one mixer the rotor speed was too slow to rapidly generate the temperature required to initiate dynamic vulcanization, in another there were difficulties with the oil injection system, problems that are not thought to be insurmountable. Dynamic vulcanization is successful in altering the blend morphology, giving a co-continuous structure with a finer texture than found in the control blends.

Most DV blends possess good extrusion performance, scoring 13 or better using the ASTM D2230 Garvey die system (a typical cable compound scores 11.5). The worst aspect of extrusion is the extrudate swell, which is generally 50–60%. Peroxide-cured DV blend vulcanizates have good physical properties, which easily exceed the specifications required by one cable manufacturer for general marine cable. The ozone resistance is good, providing the EPM is present at 50 phr or more, and the vulcanizates perform well under accelerated daylight ageing. At 100 °C, the heat ageing is good, but at 125 °C the samples suffer too much degradation for testing thus limiting use to a maximum of 60–70 °C; use of antioxidants in the formulation should improve heat ageing performance. Vulcanization using electron beam irradiation was only partly successful. A reasonable degree of crosslinking was achieved, but the test samples suffered excessive heating due to very high beam currents and were thus porous with poor ultimate properties. Irradiation under more optimized conditions should remedy this.

## 9.11 REFERENCES

1 Coran, A.Y. (1995) Vulcanization: conventional and dynamic. *Rubber Chemistry and Technology*, **68**(3), 351–375.

2 Wampler, W.A., Gerspacher, M., Yang, H.H. and O'Farrell, C.P. (1994) Dynamic vulcanization: controlling carbon black dispersion and distribution. Paper No. 52 presented at a meeting of the Rubber Division, American Chemical Society, Pittsburgh, October 1994. Abstract (1995) in *Rubber Chemistry and Technology*, **68**(1), 180.

3 Usachev, S.V., Zakharov, N.D., Kuleznev, V.N. and Vetoshkin, A.B. (1979) Aspects on the rheological properties of blends of rubbers, and their influence on the formation of the structure of the vulcanizate. *Kauchuk i Rezina*, 1979 (8), 23 translated (1980) in *International Polymer Science and Technology*, **7**(2), T48–T51.

4 Coran, A.Y. and Lee, S. (1992) High performance elastomers prepared by dynamic vulcanization. Paper No. 2 presented at a meeting of the Rubber Division, American Chemical Society, Nashville, November 3–6.

5 Tinker, A.J. (1987) Blends of natural rubber with speciality synthetic rubbers, in *Industrial Composites based on Natural Rubber – Proceedings of a UNIDO-sponsored Workshop*, MRPRA, Hertford, pp. 103–115.

6 Brown, P.S., Loadman, M.J.R. and Tinker, A.J. (1992) Applications of FT-NMR to crosslink density determinations in natural rubber blend vulcanizates. *Rubber Chemistry and Technology*, **65**(4), 744–760.

7 Technical literature of Montedison S.p.a., Milan, Italy, 1984.

8 Table F204–F205 (1980) in *The Handbook of Chemistry and Physics*, 61st edn (ed. R.C. Weast), CRC Press, Boca Raton Florida, USA. Personal communication regarding the effect of cloud cover from staff at the Meteorological Office, Bracknell, UK.

# 10

# Natural rubber/ENR-25 blends

*Sally Groves*
*Tun Abdul Razak Research Centre, MRPRA, UK*

## 10.1 INTRODUCTION

High damping rubber vulcanizates with good physical properties and a small dependence of properties on temperature are candidates for use in bushings and other bearings. Classically high damping may be achieved by either using a rubber with inherent high damping or by adding high levels of fillers and plasticizers to one which has low damping. Both approaches fail to provide the necessary combination of properties; rubbers which are highly damping also show a strong dependence of properties on temperature, whilst the use of high levels of filler and plasticizer, generally used with a low level of curatives, leads to unsatisfactory physical properties.

The aim was to develop materials with the sought characteristics by blending a highly filled and plasticized elastomer with normally compounded natural rubber (NR). It is vital that the plasticizer partitions strongly in favour of the highly filled elastomer. This is ensured by using a second elastomer which is more polar than NR and an appropriate polar plasticizer. The partitioning of the plasticizer performs several important functions. It maintains a reasonable balance between the viscosities of the two rubber phases during blending and later between the moduli of the two phases in the vulcanizate. Polar elastomers tend to have relatively high glass transition temperatures ($T_g$), and hence dependence of properties on temperature, but the plasticizer serves to reduce $T_g$.

*Blends of Natural Rubber*. Edited by Andrew J. Tinker and Kevin P. Jones.
Published in 1998 by Chapman & Hall, London. ISBN 0 412 819406

Twenty-five mole% epoxidized NR (ENR-25), is a chemically modified form of NR produced commercially from NR latex, hydrogen peroxide and formic acid. The unsaturation in one quarter of the isoprene repeat units is converted to epoxide. The rubber retains the inherent high strength of NR but has a higher polarity. ENR-25 is therefore a good candidate for the second component of the high damping blends.

The general viability of blending normally compounded NR with highly filled ENR-25 to achieve such properties has been demonstrated previously [1]. Polar plasticizers which favour the ENR-25 phase are required and a number of suitable systems have been identified [2–4]. The cure system used must be low in sulphur and highly efficient to ensure the ENR phase does not suffer from poor ageing [5].

The type and loading of carbon black filler will have a profound effect on the properties of NR/ENR-25 blends and for this reason a study of the most suitable black type and loading required in each of the blend phases was undertaken.

## 10.2 DESIGNED EXPERIMENTS

The method used to optimize filler parameters was factorial design analysis [6, 7]. This method enables the effects of more than one factor at a time to be studied (and more importantly the combined effect of and interactions between variables). The method involves varying several factors simultaneously. In the case of carbon black there are three variables: surface area, structure and loading. The filler parameters in both the NR and ENR phases of the blend were investigated in two separate designed experiments. The filler variables were kept constant in one masterbatch whilst being systematically varied in the other. The experimental design utilized was based on the central rotatable composite design. In an orthogonal design, five levels are assigned to each variable which are rotatable around a design centre point. Carbon black, however, is not available commercially at all the values of structure and surface area which are required for an orthogonal design, and the actual values in design units used in this study are shown in brackets in Table 10.1.

Compounds were prepared in an internal mixer. The mixing procedure followed is important, as it is essential to mix consistently to avoid the introduction of any uncontrolled variables to the design. NR and ENR masterbatches were prepared separately, cross-blended at a 1:1 volume ratio and then finalized by adding curatives. Compounds were vulcanized to rheometer $t_{max}$ at 150 °C and physical tests performed on the resultant vulcanizates. Regression analysis of physical test data was performed in order to relate the measured properties to the three filler parameters. Response equations are generated which have the following form:

**Table 10.1** Values of the filler parameters used in the designed experiments

| *Nominal design units* | *−1.68* | *−1* | *0* | *1* | *+1.68* |
|---|---|---|---|---|---|
| Surface area, CTAB, $m^2/g$ | 28.0 (−1.53) | 39 (−1.22) | 83 | 119 | 128 (+1.25) |
| Structure, DBPA, ml/100 g | 72 (−1.3) | 90 (−0.52) | 102 | 125 | 113 (+1.78) |
| Loading in ENR, phr[a] | 59.8 | 70 | 85 | 100 | 110.2 |
| Loading in the NR, phr[b] | 13.2 | 20 | 30 | 50 | 46.8 |

[a] In this experiment, the NR masterbatch was loaded with 30 phr N330 (CTAB 83 $m^2/g$, DBPA 102 ml/100 g).
[b] In this experiment, the ENR-25 masterbatch was loaded with 99 phr N326 (CTAB 83 $m^2/g$, DBPA 72 ml/100 g).

$$\begin{aligned}\text{Property value} = {} & C_0 + C_x V_x + C_y V_y + C_z V_z + C_{xx} V_x V_x + C_{yy} V_y V_y \\ & + C_{zz} V_z V_z + C_{xy} V_x V_y + C_{xz} V_x V_z \\ & + C_{yz} V_y V_z + C_{xyz} V_x V_y V_z\end{aligned}$$

where $V$ corresponds to the design coordinate for each variable, $x$, $y$ or $z$, that is surface area, structure and loading respectively. The response equations for both masterbatch designs were modified by using the *t*-test [7] to exclude non-valid terms and are contained in the Appendix at the end of this chapter.

Contour plots may be generated from response equations and are an indication of the change in any one particular property with change in filler parameters. Figure 10.1 is such a contour plot which represents the dependence of blend loss angle on filler surface area and loading in the ENR-25 masterbatch. The axes are in design units and the third variable (filler structure) is fixed at the centre point of the design.

The figure indicates that the highest blend loss angles are achieved at high loadings and intermediate to high surface areas of filler in the ENR-25 phase. Figure 10.2 shows the dependence of blend compression set (one day at 70 °C) on filler surface area and structure in the ENR-25 masterbatch with the loading set at the centre point of the design.

The figure indicates that low compression set values in the blend are achieved by loading the ENR-25 phase with fillers having low surface areas and low structures. The two figures illustrate a common dilemma in which there is a conflict in choice of the most suitable filler to achieve optimum properties. The most desirable situation would be to have both high damping and low compression set, but fillers which cause an in-

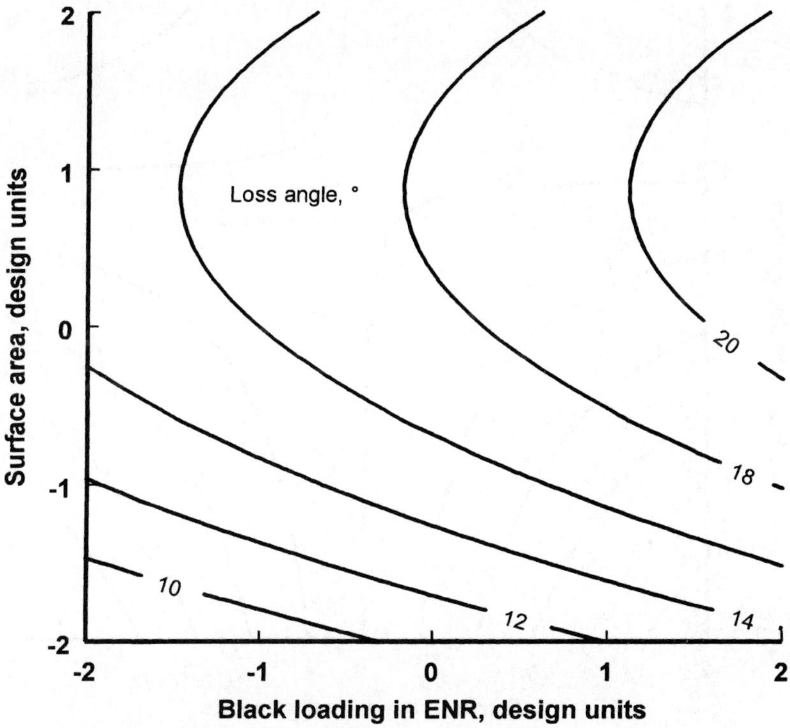


**Figure 10.1** Dependence of NR/ENR-25 loss angle (5% strain, 5 Hz, ambient) on filler surface area and loading in the ENR-25 phase.

crease in damping also increase compression set. The use of desirability functions allows the optimization of vulcanizate properties taking as many factors as are deemed relevant into account [8]. Zero desirability corresponds to an unacceptable property value whereas unity corresponds to a property value which either meets or surpasses the target set. Intermediate desirability values correspond to the degree of approximation to the most desirable value. To identify an optimum or 'preferred' formulation, individual desirabilities for the key properties are estimated from the appropriate response equations and are combined to produce an overall composite desirability function (D) [8].

The key properties used to optimize the properties of NR/ENR-25 blends were damping, compression set, tensile strength and hardness. The analysis indicated that the best compromise is achieved by loading the ENR-25 phase with between 90 and 100 phr of N326 black and the NR phase with between 30 to 40 phr of the same grade of carbon black. The loss angles attainable from these compounds are in the range 19–21° at 5% strain, 5 Hz and ambient temperatures. Masterbatch formulations

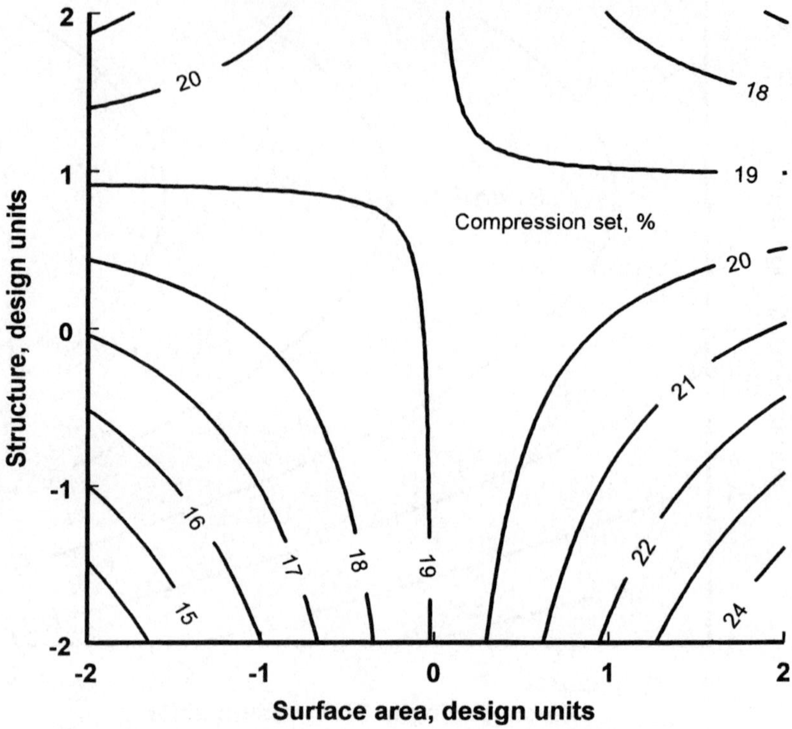


**Figure 10.2** Dependence of NR/ENR-25 compression set (one day at 70 °C) on filler surface area and structure in the ENR-25 phase.

for a typical NR/ENR-25 blend with such high loss angles are shown in Table 10.2.

Properties of a blend prepared to the formulation in Table 10.2 are shown in Table 10.3, which also contains data for a control compound.

**Table 10.2** Typical NR and ENR-25 masterbatch formulations[a]

| | | |
|---|---|---|
| SMR CV60 | 100 | – |
| ENR-25[b] | – | 100 |
| Calcium stearate | – | 3 |
| N326 | 30 | 100 |
| Diolplate 7017[c] | 6 | 34 |
| TBEP | – | 12 |

[a] Both masterbatches also contain 5 phr zinc oxide, 2 phr stearic acid, and 2 phr TMQ. Crossblended at a 1:1 volume ratio and then finalized with an EV cure system
[b] Epoxyprene-25, Kumpulan Guthrie Berhad, Malaysia
[c] A polymeric ester, Kemira Polymers, Stockport, Cheshire, UK

**Table 10.3** Properties of damping rubbers

| *Elastomer base* | *NR [9]* | *NR/ENR-25* |
|---|---|---|
| Hardness, IRHD | 66 | 49 |
| Tensile strength, MPa | 25 | 25 |
| Compression set, 1 d/70 °C, % | 18 | 16 |
| Loss angle, 5% strain, 5 Hz, ° | 13 | 19 |
| G*, 5% strain, 5 Hz, MPa | 2.3 | 2.2 |
| $G_{stat}$, 0–50%, MPa | 0.89 | 0.58 |

The latter is a single polymer NR vulcanizate compounded with the same overall level of filler as used in the NR/ENR-25 blend [9].

The table illustrates that compression set is slightly lower and the damping much higher in the blend compared with the single polymer NR vulcanizate. NR vulcanizates with similar levels of damping to the blend have very high compression set values and are much stiffer compounds [9]. The approach used in this work has therefore been effective in achieving high damping with good physical properties.

## 10.3 PLASTICIZER INVESTIGATIONS

One method of incorporating high levels of plasticizer is to use a combination of two or more plasticizers to enable a greater overall amount to be added before bleeding occurs. Two suitable plasticizers for NR/ENR-25 blends have been identified [3,4] – Diolpate 7107, a commercial blend of polymeric esters, and TBEP, tri(butoxyethyl)phosphate $(C_4H_9OC_2H_4O)_3PO$. This system has been shown to be very effective at reducing the $T_g$ of gum ENR-50 [2]. In the early stages of the work, plasticizer bleeding was encountered from the filled vulcanizates and it was necessary to determine partition coefficients for the filled ENR-25 system as described in Chapter 11. In an attempt to improve the low temperature properties of the blend by further reducing the $T_g$, a number of other plasticizer systems were investigated. DBEEA (di-(butoxyethoxyethyl)adipate) and TBEP were studied in typical blends at equivalent levels to the prevailing Diolpate 7017/TBEP system. The physical properties of vulcanizates containing the former plasticizers are similar to those of vulcanizates containing the latter. The temperature dependence of the dynamic properties of the three systems is, however, quite different. The temperature dependence of damping measured at 5 Hz and 5% strain in the range – 40 °C to 100 °C for the three differently plasticized, but otherwise comparable, vulcanizates is illustrated in Figure 10.3.

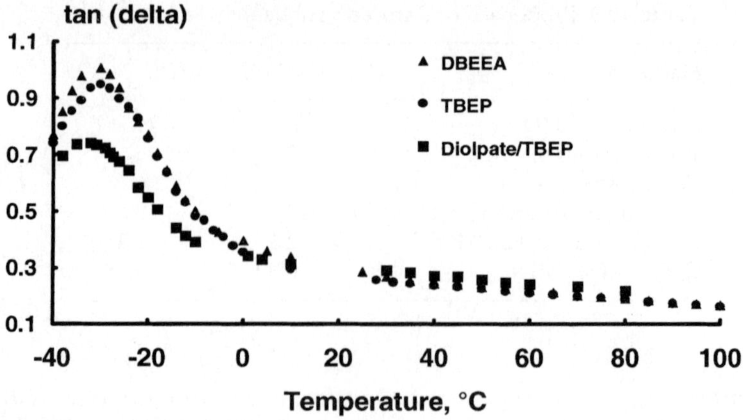


**Figure 10.3** Dependence of blend loss angle (5% strain, 5 Hz) on temperature for three differently plasticized NR/ENR-25 vulcanizates.

At ambient temperatures the loss angles are similar for all three vulcanizates. However, as the temperature is decreased, the increases in damping are more pronounced for the vulcanizates containing the single plasticizers than for that containing the dual system. All three vulcanizates show similar $T_g$s for the ENR-25 phase, as indicated by the near coincidence of temperatures at which the loss maximum is observed. Enhanced damping is beneficial for the applications for which these materials are being developed, but the three vulcanizates also show quite different dynamic stiffness dependence with temperature as depicted in Figure 10.4. The figure indicates that the dual plasticizer system gives a

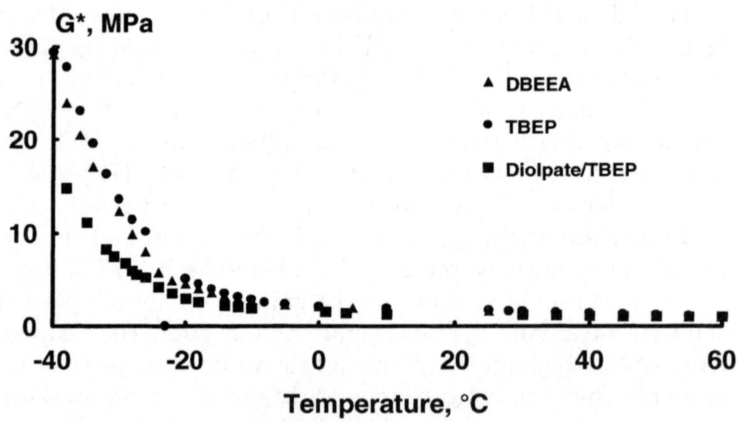


**Figure 10.4** Dependence of dynamic stiffness (5% strain, 5 Hz) on temperature for three differently plasticized NR/ENR-25 vulcanizates.

much smaller dependence of modulus on temperature than the other two plasticizers, and it is therefore preferable in terms of low temperature performance.

## 10.4 CURE SYSTEMS

The choice of cure system for these blend materials is dominated by the presence of the ENR-25 phase. ENR-25 is prone to degradation by sulphur acids resulting from network maturation on oxidative ageing, and it is vital that low sulphur levels are used in the cure system as has been shown for ENR-50 [10]. ENR-25 is also a more polar polymer than NR, and it is probable that curatives in the blend will migrate towards the ENR-25 phase. This will result in a higher crosslink density in the ENR-25 phase of the blend, evidence of which is given in Chapter 12. In principle, the lower the crosslink density in the ENR-25 phase the higher the damping. Relatively high levels in the NR phase should ensure good physical properties.

Cure systems with accelerators less liable to prefer the ENR-25 phase were selected in an attempt to reduce the crosslink density in this phase. The properties of filled compounds prepared in accordance with masterbatch formulations given in Table 10.2 and vulcanized with a variety of cure systems are presented in Table 10.4. Crosslink densities in the individual phases of gum vulcanizates cured with the same systems, as estimated by swollen-state NMR spectroscopy, are also included in Table 10.4.

The crosslink density values are consistent with physical properties in that they may be related to properties such as hardness and tensile

**Table 10.4** Effect of cure system on NR/ENR-25 blends; physical crosslink densities for gum vulcanizates, properties for filled vulcanizates

| *Property* | *EV cure system (phr)* | | | |
|---|---|---|---|---|
| | *S/CBS* | *S/TBBS* | *S/TBBS/ODIP*[a] | *TMTD/MBTS* |
| NR, $n_{phys}$, mol/m$^3$ | 22 | 25 | 23 | 26 |
| ENR-25, $n_{phys}$, mol/m$^3$ | 36 | 40 | 17 | 32 |
| Hardness, IRHD | 49 | 51 | 45 | 53 |
| Tensile strength, MPa | 22 | 23 | 21 | 22 |
| Comp. set, 1 d/70 °C, % | 30 | 25 | 28 | 24 |
| Loss angle, 5 Hz, ° | 21 | 21 | 19.5 | 19.5 |

[a] N,N'-dioctadecyl, N,N'-diisopropyl thiuramdisulphide

strength. However, it was not possible to relate crosslink densities of the ENR-25 phase alone to damping, rather that the lower crosslink densities for NR may be contributing to the damping. There is also a correlation between compression set values and total crosslink densities. The cure system chosen as most favourable overall in terms of physical properties, ageing resistance and damping properties was based on sulphur and N-*t*-butylbenzothiazole-2-sulphenamide (TBBS). Work described in Chapter 12 determined the most suitable levels to be 0.4 phr S and 3 phr TBBS.

## 10.5 DYNAMIC:STATIC MODULUS RATIOS

During the course of the work, it became apparent that the dynamic: static modulus ratios (G*, 5% strain, 5 Hz/$G_{stat}$, 5% strain, 30%/min) of these blends do not behave as expected from data for NR vulcanizates [11]. The blend vulcanizates have much lower G*:$G_{stat}$ ratios than expected for such high levels of damping, as demonstrated in Figure 10.5 which shows the values for three typical blends in comparison with the general trend line for NR vulcanizates.

This is of importance in engineering applications in which it is often beneficial to have similar properties under rest and working conditions. The exact values of the dynamic:static modulus ratio varies with formulation and is considered to be related to differing crosslink densities within the phases of the blends.

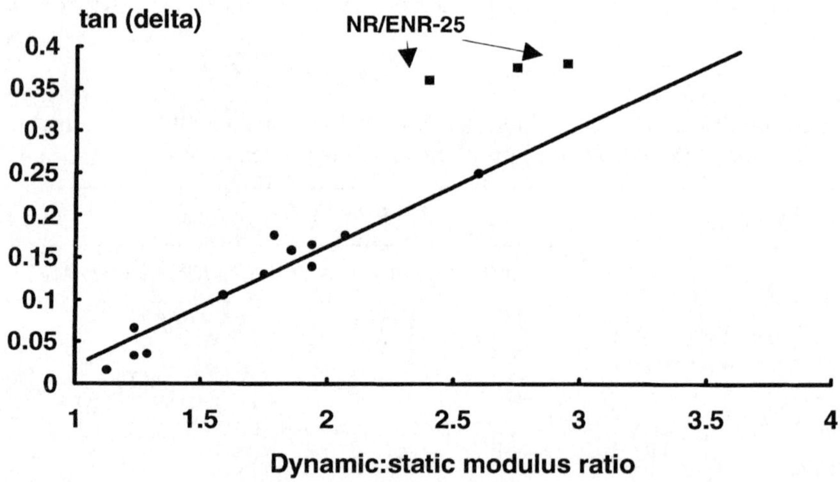


**Figure 10.5** Dependence of loss angle on G* (5% strain, 5 Hz):$G_{stat}$ (30% strain/min) ratio for NR and NR/ENR-25 vulcanizates.

## 10.6 TRI-BLENDS

It has been shown previously [12] that replacement of 10% by weight of NR with high vinyl polybutadiene (Hv-BR) significantly reduces the rate of low temperature crystallization of NR. The NR phase of blends described here is prone to high rates of crystallization because of the low crosslink densities in the NR phase and highly efficient cure system used [13]. Tri-blends in which 10% of the polymer in the NR phase was replaced with Hv-BR, have been produced in an identical manner to the NR/ENR-25 bi-blends. There is little difference between the physical properties of the two materials at ambient temperature as shown in Chapter 12. However the rate of increase in hardness at −26 °C of the tri-blend is lower than that of the bi-blend (Figure 10.6). This indicates that the tri-blends have improved resistance to low temperature crystallization of the NR phase. This matter is considered further in Chapter 12.

## 10.7 CONCLUSIONS

NR/ENR-25 bi-blends and tri-blends containing Hv-BR have been shown to produce high damping vulcanizates with good physical properties. N326 carbon black is the preferred filler as it gives the best compromise in properties. Different plasticizers impart different dependences of vulcanizate properties on temperature, and a dual Diolpate/TBEP system is recommended as this minimizes changes in properties with temperature. Tri-blends are more resistant to progressive

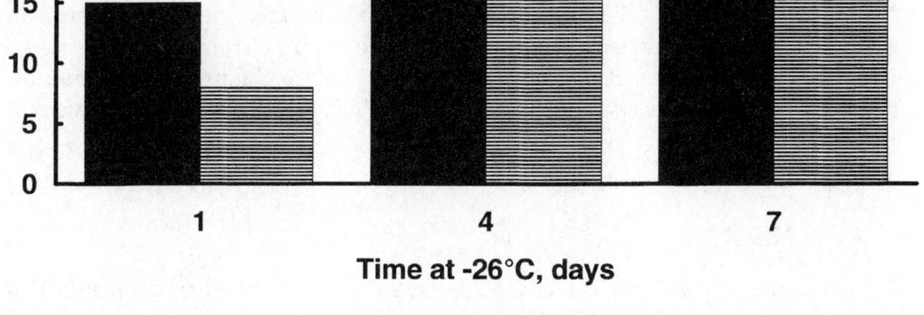


**Figure 10.6** Increase in hardness at −26 °C for NR/ENR-25 and NR/HvBR/ENR-25 blends.

stiffening at low temperatures due to crystallization of the NR component of the blend, whilst other physical properties remain comparable to those of the bi-blends. These blends in general have a significant advantage over conventional high damping NR vulcanizates in that they have much lower dynamic:static modulus ratios for materials with such high damping, and this is a very desirable property.

## 10.8 REFERENCES


1 Brown, P.S. (1996) High damping NR/ENR blends. *Journal of Natural Rubber Research*, submitted for publication.
2 Aris bin Ahmad, Tinker, A.J. and Aubrey, D.W. (1994) Absorption of ester type plasticisers by natural rubber and epoxidised natural rubber – diffusion coefficients and equilibrium uptake. *Journal of Natural Rubber Research*, **9**(1), 1–12.
3 Aris bin Ahmad, Tinker, A.J. and Aubrey, D.W. (1995) Distribution of ester plasticisers between two rubber phases – estimation of partition coefficients and comparison with theory. *Journal of Natural Rubber Research*, **10**(1), 1–13.
4 Aris bin Ahmad (1990) *The study of the partition of plasticizer and filler for highly hysteretic rubber blends*. PhD Thesis, CNAA, London.
5 Gelling, I.R. and Metherell, C. (1993) Vulcanisation systems for ENR-50. *Journal of Natural Rubber Research*, **8**(1), 37–56.
6 Freakley, P.K. (1985) *Rubber Processing and Product Organisation*. Plenum Press, New York.
7 Davies, O.L. and Goldsmith, P.L. (1972) *Statistical Methods in Research and Production*. Oliver and Boyd, Edinburgh.
8 Derringer, G.C. (1983) A statistical methodology for designing elastomer formulations to meet performance specifications. *Kautschuk und Gummi Kunststoffe*, **35**(9), 349–352.
9 MRPRA (1983) *Natural Rubber Engineering Data Sheet* No. 46, ISSN 0141–9560.
10 Gelling, I.R. and Morrison, N.J. (1985) Sulphur vulcanization and oxidative ageing of Epoxidised Natural Rubber. *Rubber Chemistry and Technology*, **58**(2), 243–257.
11 MRPRA (1983) *Natural Rubber Engineering Data Sheets*, ISSN 0141–9560.
12 Brown, P.S., Cook, S., Groves, S.A., Lewan, M.V. and Tinker, A.J. (1993) Natural rubber blends. Paper No. 127 presented at a meeting of the American Chemical Society, Rubber Division, Orlando, Florida, October 26–29.
13 Crowther, B.G., Lewis, P.M. and Metherell, C. (1988) Chapter 6, Compounding, *Natural Rubber Science and Technology* (ed A.D. Roberts) Oxford University Press, Oxford, p. 180.


## 10.9 APPENDIX

Valid response equations for ENR-25 black type where the variables $x$, $y$ and $z$ represent carbon black surface area (CTAB, $m^2/g$), structure (DPBA, ml/100 g) and loading (phr) respectively:

Mooney viscosity of ENR masterbatch – ML(1+4), 100 °C

$$= 55.24 + 11.23V_x + 9.921V_y + 14.54V_z + 4.206V_xV_z + 4.095V_yV_z$$

Rheometer torque maximum, $M_H$(dNm) – Monsanto MDR 2000 at 150 °C

$$= 13.53 + 1.103V_x + 0.3421V_y + 1.137V_z + 0.1365V_xV_x + 0.1877V_yV_y - 0.3183V_xV_y + 0.2461V_xV_z - 0.1477V_yV_z$$

Cure time, $t_{95}$ (minutes) – Monsanto MDR 2000 at 150 °C

$$= 16.1 - 1.034V_y + 0.7089V_z + 1.559V_xV_y$$

Hardness (IRHD) – BS903:A26, ISO48, ISO1818

$$= 60.27 + 2.178V_x + 2.148V_y + 3.571V_z + 0.6754V_xV_z$$

Tensile strength (MPa) – BS903:A2, ISO37

$$= 21.04 + 0.9668V_x - 0.6291V_y - 0.581V_z + 0.4045V_xV_z$$

Elongation at break (%) – BS903:A2, ISO37

$$= 489 + 12.29V_x - 61.57V_y + 10.03V_xV_z + 16.04V_yV_z$$

Compression set, 3 days at 23 °C (%) – BS903:A6, ISO815

$$= 9.027 + 1.326V_x$$

Compression set, 1 day at 70 °C (%) – BS903:A6, ISO815

$$= 19.07 + 0.9946V_y - 1.049V_xV_y$$

Crescent tear strength (N/mm) – BS903:A3, ISO34

$$= 88.91 + 8.997V_x + 4.696V_y + 7.649V_z - 9.321V_xV_x - 0.31V_yV_y - 1.749V_zV_z + 5.869V_xV_y + 1.573V_xV_z + 2.239V_yV_z$$

Loss angle, δ, at 5% strain and 5 Hz (°) – Instron 1271, double shear

$$= 17.56 + 1.648V_x + 1.546V_z - 0.9488V_xV_x$$

Valid response equations for NR black type where the variables $x$, $y$ and $z$ represent carbon black surface area (CTAB, $m^2/g$), structure (DPBA, ml/100 g) and loading (phr) respectively:

Mooney viscosity of NR masterbatch – ML(1+4), 100 °C

$$= 63.16 + 4.058V_x + 2.947V_y + 8.507V_z + 1.152V_zV_z$$

Rheometer torque rise, $M_H - M_L$ (dNm) – Monsanto MDR 2000 at 150 °C

$$= 11.67 + 0.3165V_x - 0.383V_y + 0.912V_z$$

Cure time, $t_{95}$ (minutes) – Monsanto MDR 2000 at 150 °C

$$= 14.76 - 1.495V_y + 1.305V_yV_y - 0.7061V_zV_z$$

Hardness (IRHD) – BS903:A26, ISO48, ISO1818

$= 59.87 + 0.943V_x + 3.258V_z - 0.8948V_zV_z$

Relaxed modulus, MR100 (MPa) – BS903:A2, ISO37

$= 2.039 + 0.2149V_z - 0.0910V_zV_z$

Tensile strength (MPa) – BS903:A2, ISO37

$= 20.63 + 0.9351V_x$

Compression set, 3 days at 23 °C (%) – BS903:A6, ISO815

$= 10.03 + 0.7386V_x + 0.755V_xV_y$

Compression set, 1 day at 70 °C (%) – BS903:A6, ISO815

$= 21.78 + 1.199V_zV_z$

Trouser tear strength (N/mm) – BS903:A3, ISO34

$= 13.07 + 2.144V_x - 2.104V_y$

Crescent tear strength (N/mm) – BS903:A3, ISO34

$= 83.1 + 4.87V_z$

Loss angle, δ, at 5% strain and 5 Hz (°) – Instron 1271, double shear

$= 18.5 + 0.7695V_x + 1.65V_z - 0.7606V_xV_x$

# 11

# Partition coefficients for an ester plasticizer in black-filled blends

*Jianhe Liao* and Sally Groves*[†]
**Chinese Academy of Tropical Agricultural Sciences;*
*[†]Tun Abdul Razak Research Centre, MRPRA, UK*

## 11.1 INTRODUCTION

In recent years there has been a requirement for rubber vulcanizates to combine high damping, good physical properties and a low dependence of dynamic properties on temperatures for use in applications such as automotive bushings. Prior research [1, 2] indicated that novel blends of natural rubber (NR) and 25 mole% epoxidized NR (ENR-25), in which the ENR-25 phase contains high levels of filler and plasticizer, can have an appropriate combination of high damping, good physical properties and low dependence of properties on temperature. Incorporating high levels of carbon black into the ENR-25 phase of NR/ENR-25 blends to increase hysteresis necessitates the use of high levels of plasticizer which must remain largely in the ENR-25 phase of the blend. The plasticizer serves key roles: it balances the viscosities of the two rubber masterbatches, thus improving processing and easing the formation of a fine phase morphology, it balances the moduli of the two rubber phases in the vulcanizate, and it reduces the glass transition temperature ($T_g$) of the ENR-25 to minimize the dependence of properties on temperature.

As a further method to reduce the effects of low temperatures on properties, it is possible to blend the NR phase with a small proportion of high vinyl polybutadiene (Hv-BR). This has been shown to inhibit low

*Blends of Natural Rubber*. Edited by Andrew J. Tinker and Kevin P. Jones.
Published in 1998 by Chapman & Hall, London. ISBN 0 412 819406

temperature crystallization of NR [3,4] without causing significant changes in other properties as demonstrated in Chapter 12.

In order for the plasticizer to largely remain in the ENR-25 phase after blending it must have a favourable partition coefficient ($K_d$). Data concerning plasticizer distributions and loading levels in NR/ENR-25 blends and single polymer gum vulcanizates for a range of plasticizers are available in the literature [5,6], and appropriate candidates were identified. However, work performed on black-filled NR/ENR-25 blends using the reported values produced vulcanizates which exude plasticizer after a period of weeks (Chapter 10). It was, therefore, necessary to estimate partition coefficients for the plasticizers in black-filled NR/ENR-25 and NR/Hv-BR/ENR-25 vulcanizates.

Bleeding may be due to either or both of two factors – the extent of crosslinking and the presence of carbon black. An increased level of crosslinking in the filled vulcanizates in respect of the levels in a previous study [6] may be reducing the extent to which the network can swell, and therefore, produce bleeding. Carbon black may also contribute to elastic constraints of swelling by rubber–filler interactions.

## 11.2 IDENTIFICATION OF EXUDING PLASTICIZER

The vulcanizates which showed exudate (Chapter 10) contained two different plasticizers, a polymeric ester (Diolpate 7017) and tri butoxyethyl phosphate (TBEP). These had been introduced in an attempt to achieve a greater overall level of plasticization than would be possible by using a single plasticizer. The initial problem therefore was to determine which of the plasticizers in the dual system was contained in the exudate. NR/ENR-25 blends were prepared to the formulations in Table 11.1 in a twin rotor internal mixer following the established procedure of mixing NR and ENR-25 carbon black masterbatches separately, cross-blending and then adding curatives.

Addition of powders, plasticizer and carbon black followed a strict routine for each of the mixes. The masterbatches were cross-blended at a 1:1 volume ratio and finalized with the curative levels also shown in Table 11.1. Vulcanizates were prepared by curing to rheometer $t_{95}$ at 150 °C. Exudate removed from the surface of the blend vulcanizate was dissolved in chloroform and analysed by GC–MS and NMR spectroscopy. Both techniques detected the presence of Diolpate 7017 in the exudate, but no TBEP.

## 11.3 ESTIMATION OF DIOLPATE PARTITION COEFFICIENTS

It was necessary to estimate the partition coefficients for Diolpate 7017 to determine the levels to incorporate into each masterbatch in order to

**Table 11.1** NR and ENR-25 masterbatch formulations

| | | |
|---|---|---|
| NR | 100 | – |
| ENR-25[a] | – | 100 |
| N326 carbon black | 30 | 99 |
| Diolpate 7017[b] | 6 | 34 |
| TBEP | – | 11 |
| Zinc oxide | 5 | 5 |
| Stearic acid | 2 | 2 |
| TMQ | 2 | 2 |
| Calcium stearate | – | 3 |
| Crossblended at a 1:1 volume ratio | | |
| Sulphur | 0.3 | |
| TMTD | 2.0 | |
| TBBS | 3.0 | |

[a] Epoxyprene-25, Kumpulan Guthrie Berhad, Malaysia
[b] A polymeric ester, Kemira Polymers, Stockport, Cheshire, UK

reduce bleeding in the filled blend. Table 11.2 shows the formulations of NR, NR/Hv-BR and ENR-25 vulcanizates used in the determination of $K_d$ for Diolpate 7017.

Plasticizer was introduced to the system by incorporating it at the mixing stage in the ENR-25 masterbatch. The method for determining $K_d$

**Table 11.2** Formulations of vulcanizates used in determination of $K_d$ of Diolpate 7017

| *Polymer* | *NR* | | | *ENR* | | | *NR/Hv-BR* | | |
|---|---|---|---|---|---|---|---|---|---|
| *Mix Ref.* | *1* | *2* | *3* | *1* | *2* | *3* | *1* | *2* | *3* |
| NR | 100 | 100 | 100 | – | – | – | 90 | 90 | 90 |
| ENR-25 | – | – | – | 100 | 100 | 100 | – | – | – |
| Hv-BR[a] | – | – | – | – | – | – | 10 | 10 | 10 |
| N326 carbon black | 30 | 30 | 30 | 100 | 100 | 100 | 30 | 30 | 30 |
| Diolpate 7017 | – | – | – | 40 | 40 | 40 | – | – | – |
| Zinc oxide | 5 | 5 | 5 | 5 | 5 | 5 | 5 | 5 | 5 |
| Stearic acid | 2 | 2 | 2 | 2 | 2 | 2 | 2 | 2 | 2 |
| TMQ | 2 | 2 | 2 | 2 | 2 | 2 | 2 | 2 | 2 |
| Calcium stearate | – | – | – | 3 | 3 | 3 | – | – | – |
| Sulphur | 0.6 | 1.3 | 2.4 | 0.6 | 1.3 | 2.4 | 0.6 | 1.3 | 2.4 |
| MBS | 0.6 | 1.3 | 2.4 | 0.6 | 1.3 | 2.4 | 0.6 | 1.3 | 2.4 |

[a] Buna VI 1979, Bayer, France

involved plying together 2 mm thick squares of NR (or NR/Hv-BR) and ENR-25 vulcanizates and then maintaining surface contact between them by encasing them in foil following the published method [6]. All combinations of the NR or NR/Hv-BR vulcanizates with ENR-25 vulcanizate were considered. The change in mass of each vulcanizate with time was recorded. When a state of equilibrium is achieved, then

$$K_d = \frac{[\text{plast}_{NR}]}{[\text{plast}_{ENR}]}$$

where $[\text{plast}_{NR}]$ and $[\text{plast}_{ENR}]$ are the concentrations of plasticizer in the NR and ENR vulcanizates respectively [7]. Concentrations were calculated on the basis of only the polymer and plasticizer in the vulcanizates by the method used in the reference work [1, 6] and the resultant $K_d$ values are shown in Table 11.3. Since NR and Hv-BR are miscible [3], the NR/Hv-BR blend is treated as a single phase.

The values were all less than unity and confirmed that Diolpate 7017 will partition in favour of the ENR-25 phase in black-filled NR/ENR-25 and NR/Hv-BR/ENR-25 blends. The data are represented as contour plots for NR/ENR-25 blends in Figure 11.1 and NR/Hv-BR/ENR-25 blends in Figure 11.2. The plots indicate that increasing the NR or NR/Hv-BR crosslink density decreases $K_d$ and increasing the ENR-25 crosslink density increases $K_d$. These observations are consistent with the previous work concerning gum vulcanizates [6]. The effect of crosslink density on $K_d$ is greater in the NR vulcanizate than the ENR-25. This observation may imply that the effect of elastic constraint by crosslinking is proportionally less in ENR-25 where the polymer–plasticizer interaction is much greater.

**Table 11.3** Partition coefficients of NR/ENR-25 and (NR/Hv-BR)/ENR-25 calculated four months after plying

| *Curative, S phr* | | $K_d$ | |
|---|---|---|---|
| *NR(NR/Hv-BR)* | *ENR-25* | *NR/ENR-25* | *(NR/Hv-BR)/ENR-25* |
| 0.6 | 0.6 | 0.0452 | 0.0449 |
| 0.6 | 1.3 | 0.0461 | 0.0465 |
| 0.6 | 2.4 | 0.0458 | 0.0497 |
| 1.3 | 0.6 | 0.0429 | 0.0461 |
| 1.3 | 1.3 | 0.0436 | 0.0458 |
| 1.3 | 2.4 | 0.0437 | 0.0488 |
| 2.4 | 0.6 | 0.0370 | 0.0412 |
| 2.4 | 1.3 | 0.0378 | 0.0423 |
| 2.4 | 2.4 | 0.0386 | 0.0437 |

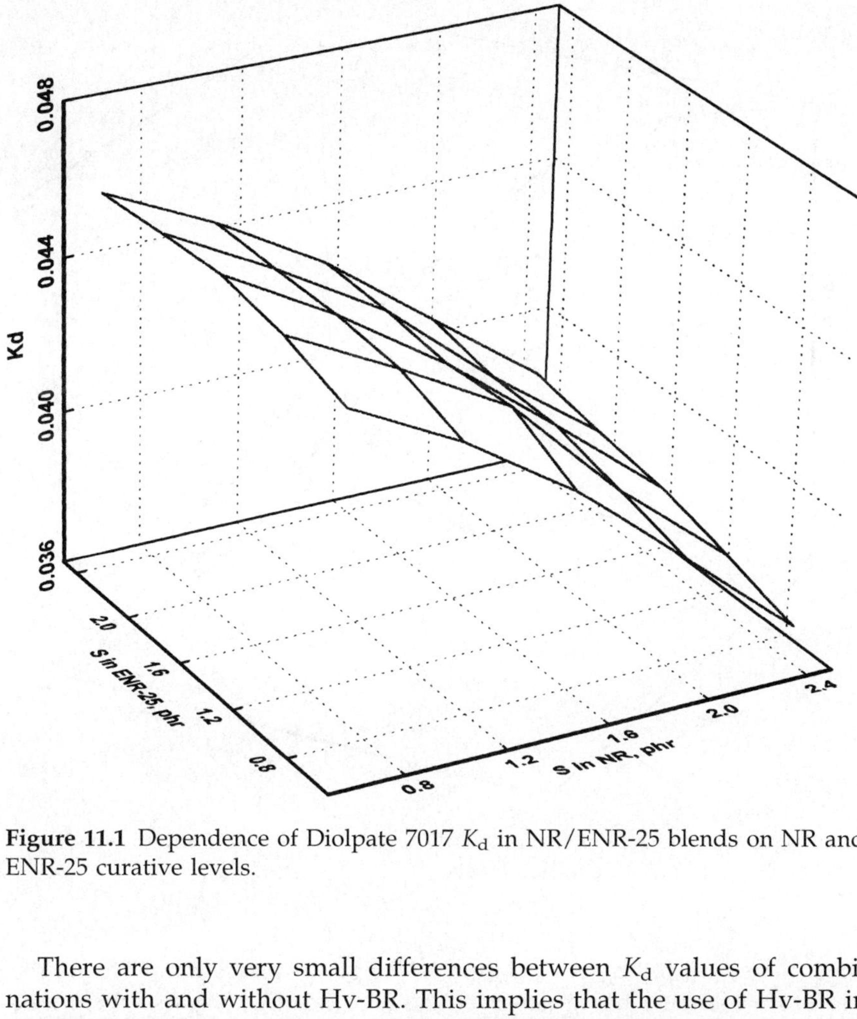


**Figure 11.1** Dependence of Diolpate 7017 $K_d$ in NR/ENR-25 blends on NR and ENR-25 curative levels.

There are only very small differences between $K_d$ values of combinations with and without Hv-BR. This implies that the use of Hv-BR in NR/ENR-25 blends will not significantly affect whether the plasticizer will bleed or not.

The $K_d$ value reported for an unfilled NR/ENR-25 combination with a cure system close to that of sample NR2/ENR2 is 0.262[6], the $K_d$ of the black-filled NR2/ENR2 is 0.0436, less than a fifth of the value for gum vulcanizates. The filled materials therefore have a significantly greater proportion of plasticizer in the ENR-25 than expected from the work on gum vulcanizates. This may also be due to the effects of constraint on elasticity due to rubber–filler interactions being greater in the NR which is not favoured by the plasticizer.

The partition coefficients observed here for black-filled vulcanizates demonstrate that it is not possible to predict partition of plasticizer in

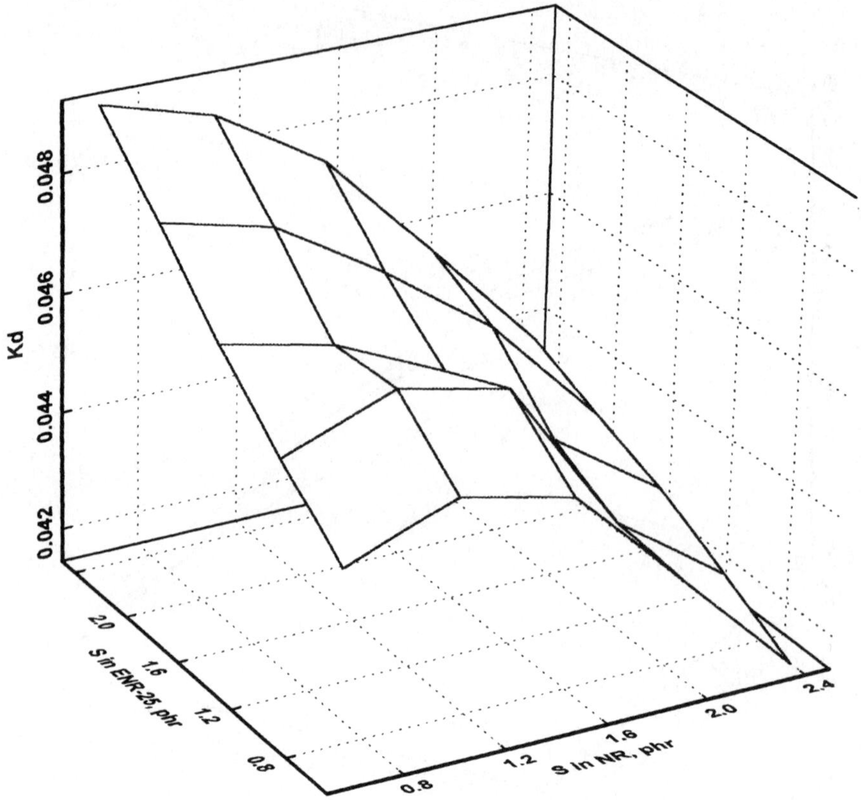


**Figure 11.2** Dependence of Diolpate 7017 $K_d$ in NR/Hv-BR/ENR-25 blends on NR/Hv-BR and ENR-25 curative levels.

filled blends on the basis of estimates for gum vulcanizates. This is the reason for bleeding being observed from black-filled blends when none was apparent from comparable gum blends. It would be desirable to determine the level at which bleeding no longer occurs in the filled blends and the effect of reducing plasticizer levels on compound properties. It is also evident that changing the volume ratio of masterbatches used in the blend will affect bleeding and vulcanizate properties. A two factor designed experiment to investigate the effects of varying Diolpate level and masterbatch ratio is described in Chapter 13.

## 11.4 CONCLUSION

The estimation of $K_d$ values for Diolpate 7017 in filled NR/ENR-25 or NR/Hv-BR/ENR-25 blends indicates that increasing the NR or NR/Hv-

BR crosslink density decreases $K_d$; with increasing ENR-25 crosslink density, $K_d$ increases. Carbon black-filled vulcanizates show significantly reduced partition coefficients relative to comparable gum vulcanizates.

## 11.5 REFERENCES

1 Aris bin Ahmad (1990) The study of the partition of plasticizer and filler for highly hysteretic rubber blends. PhD Thesis, CNAA, London.
2 Brown, P.S. (1996) High damping NR/ENR blends. *Journal of Natural Rubber Research*, submitted for publication.
3 Roland, C.M. (1988) Interactions in a miscible polymer blend. *Rubber Chemistry and Technology*, **61**(5), 866–878.
4 Brown, P.S., Cook, S., Groves, S.A., Lewan, M.V. and Tinker, A.J. (1993) Natural rubber blends. Paper No. 127 presented at a meeting of the American Chemical Society, Rubber Division, Orlando, Florida, October 26–29.
5 Aris bin Ahmad, Tinker, A.J. and Aubrey, D.W. (1994) Absorption of ester type plasticizers by natural rubber and epoxidised natural rubber – diffusion coefficients and equilibrium uptake. *Journal of Natural Rubber Research*, **9**(1), 1–12.
6 Aris bin Ahmad, Tinker, A.J. and Aubrey, D.W. (1995) Distribution of ester plasticizers between two rubber phases – estimation of partition coefficients and comparison with theory. *Journal of Natural Rubber Research*, **10**(1), 1–13.
7 Reilly, J. and Rove, W.N. (1948) Partition coefficients, in *Physico-Chemical Methods, Vol. 3*, Methuen and Co. Ltd., London.

# 12

# Improving resistance to low temperature crystallization in NR/ENR-25 blends

*Benny George*, Jianhe Liao† and Sally Groves§*
**Rubber Research Institute of India;*
*†Chinese Academy of Tropical Agricultural Sciences;*
*§Tun Abdul Razak Research Centre, MRPRA, UK*

## 12.1 INTRODUCTION

The automobile rubber products manufacturing industry has expressed a long-standing need for high damping rubbers which combine good physical properties with a low dependence of dynamic properties on temperature. Elastomers with inherently high damping have a high dependence of properties on temperature because the damping derives from relatively high glass transition temperatures ($T_g$). An alternative approach of using natural rubber (NR) with high levels of carbon black and oil but relatively low crosslink densities limits the degree of damping attainable and gives rather poor physical properties.

Previous work has demonstrated the advantage of using a blend of normally filled NR with highly filled and plasticized 25 mole% epoxidized NR (ENR-25) [1]. The NR component provides good physical properties whilst the ENR-25 phase provides high damping. These blends are required to provide good performance at both high and low temperatures. The use of efficient (EV) or semi-EV sulphur cure systems, in which the levels of accelerator are greater than those of sulphur, and appropriate antioxidants should provide resistance to change in prop-

*Blends of Natural Rubber*. Edited by Andrew J. Tinker and Kevin P. Jones.
Published in 1998 by Chapman & Hall, London. ISBN 0 412 819406

erties at elevated temperatures [2]. However, low temperature properties of NR/ENR-25 blends can be affected by the stiffening of the NR phase due to crystallization. These blends are particularly prone to such stiffening, because the polarity of the ENR phase tends to attract the curatives to itself and leaves the NR with a low crosslink density (Chapter 10). ENR is not so susceptible to crystallization at low temperatures due to the inhibiting effects of the epoxide groups [3]. One method to overcome NR crystallization is to use a conventional sulphur vulcanization system, which imparts chain modification in NR. However this conflicts with the requirement of low sulphur levels in the ENR phase, which is prone to attack by sulphur acids [4, 5].

It had been reported that the incorporation of small amounts of high vinyl polybutadiene (Hv-BR) in NR increases the inhibition time for crystallization and decreases the rate of subsequent crystallization [6]. Therefore, the effect of incorporating Hv-BR into the NR phase of NR/ENR-25 blends as a means towards reducing low temperature crystallization was examined.

For a given cure system, the degree to which separate elastomers in a blend are vulcanized will have a profound effect on both damping and physical properties. The ideal situation for the blend described herein would be a moderate crosslink density in the NR achieved by a semi-EV cure system and a lower crosslink density in the ENR-25 achieved by an EV cure system, thereby ensuring high damping and good ageing resistance. Because the ENR is polar it is difficult to achieve a lower crosslink density in this phase because of the tendency of curatives to preferentially locate in the more polar elastomer of a blend (Chapter 5). However, earlier work had shown that a S/TBBS cure system gives satisfactory properties (Chapter 10). The approach described herein used factorial design methods for determining the S/TBBS level which provides optimum properties.

## 12.2 EFFECT OF INCORPORATING HV-BR INTO NR/ENR-25 BLENDS

The miscibility of NR and Hv-BR has been established in previous work [6, 7]. The closely related work [6] indicated that a Hv-BR content of 10% is sufficient to significantly inhibit the rate of crystallization of NR with little effect on physical properties. In this work it was decided to restrict the level of Hv-BR in the NR phase of NR/ENR-25 blends to 10% for this reason.

Blends of NR/Hv-BR with ENR-25 were studied by Dynamic Mechanical Thermal Analysis (DMTA) over the temperature range −100°–0 °C in order to confirm that the blend consisted of only two separate phases. Figure 12.1 shows the DMTA thermogram of a NR/Hv-BR/

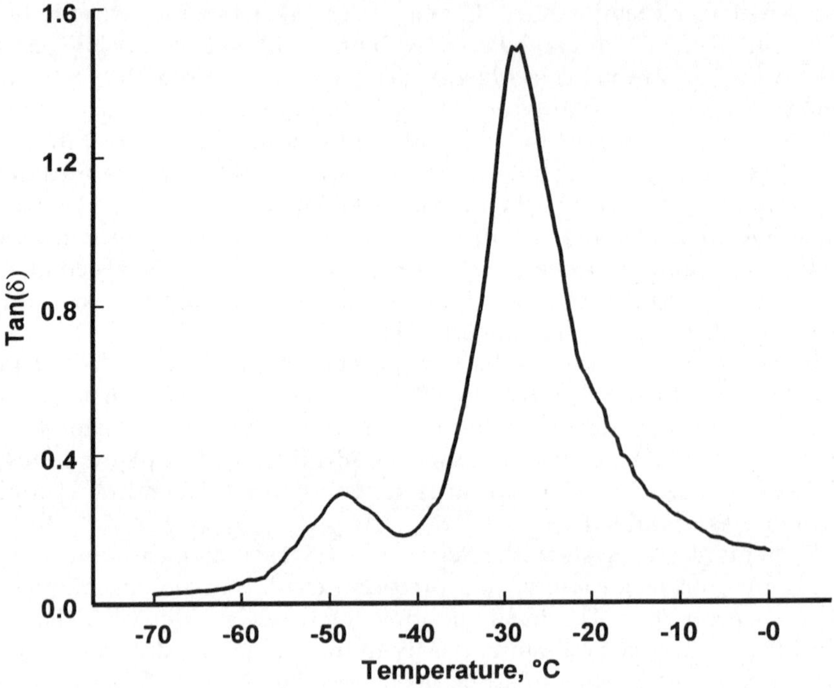


**Figure 12.1** DMTA thermogram of a 45:5:50 NR/Hv-BR/ENR-25 gum vulcanizate at 1 Hz.

ENR-25 gum vulcanizate prepared in the ratio 45:5:50. A maximum in the loss peak corresponds to a distinct $T_g$. It is clear from the figure that NR and Hv-BR form a miscible phase represented by a single $T_g$ at −48 °C. ENR-25 exhibits a separate $T_g$ at −29 °C. A separate Hv-BR phase would be manifested as a further loss maximum between −10 °C and 0 °C.

Having established the miscibility of this grade of Hv-BR with NR, filled tri-blend vulcanizates were prepared with the aim of investigating the low temperature crystallization of the NR/Hv-BR phase. Vulcanizates were prepared according to the formulations listed in Table 12.1 by first mixing separate NR/Hv-BR and ENR-25 masterbatches in an internal mixer then cross-blending these masterbatches and adding curatives in a single mix cycle.

Samples were cured to rheometer $t_{95}$ at 150 °C. The degree of crystallization in the NR/Hv-BR phase of tri-blends after periods in an unstrained state at −26 °C was determined by differential scanning calorimetry (DSC). The rate of crystallization in NR is at a maximum at this temperature [8]. The results of the DSC study are shown in

**Table 12.1** Formulations of masterbatches used in preparation of blends containing Hv-BR, phr[a]

| | | | |
|---|---|---|---|
| SMR CV | 100 | 90 | – |
| Hv-BR (VI-1979)[b] | – | 10 | – |
| ENR-25[c] | – | – | 100 |
| N326 carbon black | 30 | 30 | 100 |
| Dioplate 7017[d] | 6 | 6 | 30 |
| TBEP | – | – | 12 |
| Zinc oxide | 5 | 5 | 5 |
| Stearic acid | 2 | 2 | 2 |
| TMQ | 2 | 2 | 2 |
| Calcium stearate | – | – | 3 |

[a] Subsequently crossblended at a 1:1 volume ratio, finalized with 0.3 phr S, 2 phr TMTD and 3 phr TBBS
[b] Buna VI 1979 (70 ± 3% 1,2, content), Bayer, France
[c] Epoxyprene-25, Kumpulan Guthrie Berhad, Malaysia
[d] A polymeric ester, Kemira Polymers, Stockport, Cheshire, UK

Figure 12.2 and indicate that the rate of crystallization in NR/Hv-BR and NR/Hv-BR/ENR-25 blends is slower than that of comparable NR and NR/ENR control vulcanizates.

In addition, the DSC study also provides insight into the crosslink density of the NR phase in the blends. Increased crosslink densities are

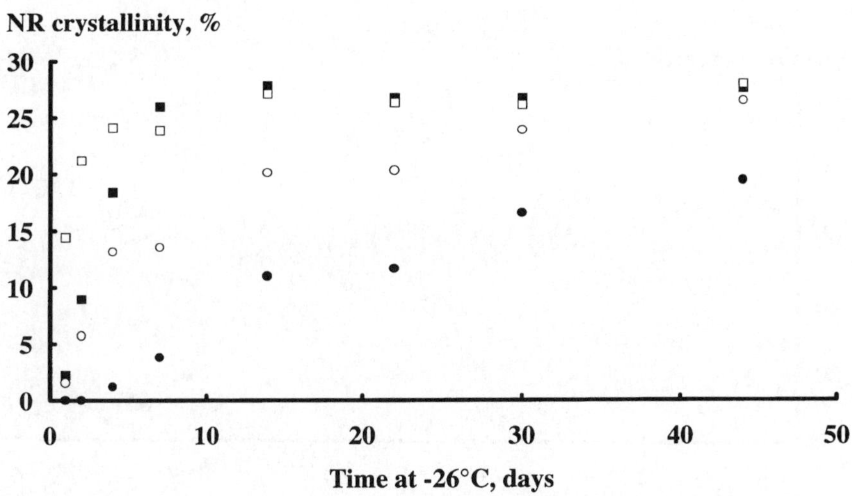


**Figure 12.2** Comparison of the degree of crystallization in the NR (NR/Hv-BR) phase of NR (■), NR/Hv-BR (●), NR/ENR-25 (□) and NR/Hv-BR/ENR-25 (○) vulcanizates at −26 °C.

known to reduce crystallization rates [8]. The data imply faster rates in the NR/ENR and NR/Hv-BR/ENR blends with respect to a single polymer NR vulcanizate. This indicates that the crosslink density in the NR phase of the NR(NR/Hv-BR)/ENR-25 blends is lower than that of the control NR single polymer vulcanizate, and reinforces the suggestion that ENR tends to scavenge curatives in this blend system.

The susceptibility of these vulcanizates to low temperature crystallization in a strained state has been determined by measuring the loss of stress associated with the onset of crystallization. Strained samples were kept at −26 °C over a time period similar to the DSC study. The results of the study are shown in Figure 12.3 and confirm that crystallization of NR/Hv-BR and NR/Hv-BR/ENR-25 blends is slower than the corresponding vulcanizates not containing Hv-BR.

In order to assess the effect of incorporating Hv-BR into NR/ENR-25 blends on properties, both NR/ENR-25 and NR/Hv-BR/ENR-25 vulcanizates were prepared to the masterbatch formulations in Table 12.1 but cured with 0.3 phr sulphur and 4 phr TBBS. Test results shown in Table 12.2 indicate that the properties of the blend containing Hv-BR are similar to those of the binary blend. The lower tear strength of the tri-blend may be attributed, at least in part, to a higher modulus apparently due to higher crosslink density.

Ozone resistance tests performed at 40 °C, 20% strain and 50 pphm ozone showed a substantial reduction in cracks (by 70%) in the tri-blend.

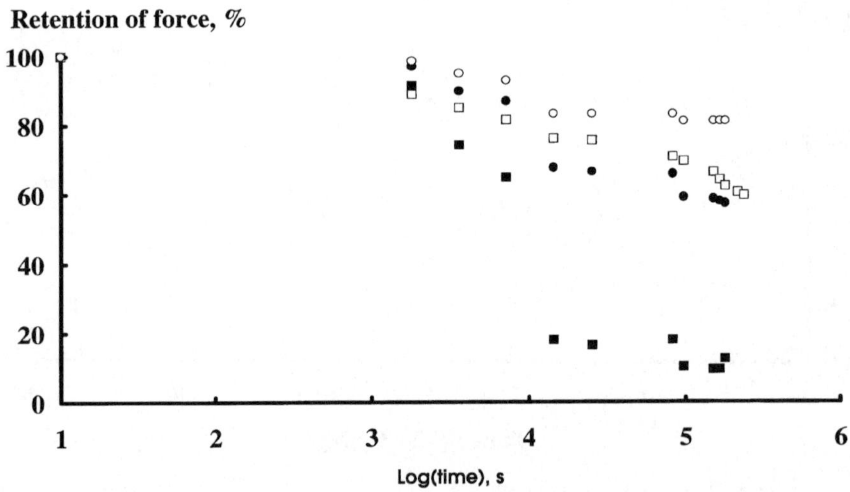


**Figure 12.3** Stress relaxation of NR (■), NR/Hv-BR (●), NR/ENR-25 (□) and NR/Hv-BR/ENR-25 (○) vulcanizates at −26 °C.

**Table 12.2** Comparison of the properties of NR/ENR-25 and NR/Hv-BR/ENR-25 vulcanizates

| *Property* | *NR/ENR-25* | *NR/HV-Br/ENR-25* |
|---|---|---|
| Mooney viscosity[a] | 48 | 49 |
| $t_{95}$, min[b] | 18.5 | 15.0 |
| $M_H$–$M_L$, dNm | 12.6 | 16.3 |
| Hardness, IRHD | 60 | 64 |
| Tensile strength, MPa | 20.1 | 20.6 |
| Elongation at break, % | 450 | 450 |
| MR100, MPa | 1.9 | 2.4 |
| Crescent tear strength, N/mm | 71 | 44 |
| Compression set, 1 d/70 °C, % | 21 | 23 |
| Compression set, 3 d/23 °C, % | 11 | 11 |
| Loss angle, 5 Hz/5%, ° | 18.5 | 17.8 |

[a] ML(1+4) 100 °C
[b] Monsanto MDR2000E rheometer at 150 °C

## 12.3 CURATIVE LEVEL OPTIMIZATION

Previous work had shown that a number of vulcanization systems give good physical properties and high loss angles for NR/Hv-BR/ENR-25 blends (Chapter 10). All were derived from formulations known to be suitable for ENR-50 vulcanizates [4, 5]. A S/TBBS system was found most appropriate, but it is possible to further improve properties by optimizing the sulphur/accelerator level. A two-level factorial designed experiment [9, 10] was performed to investigate the widest range of S/TBBS levels within an EV to semi-EV range. The masterbatch formulations are shown in Table 12.3 and the levels of sulphur and TBBS used in the design are given in Table 12.4 which also includes compound

**Table 12.3** Masterbatch formulations, phr

| | | |
|---|---|---|
| SMR CV | 90 | – |
| Hv-BR (VI 1979) | 10 | – |
| ENR-25 | – | 100 |
| Zinc oxide | 5 | 5 |
| Stearic acid | 2 | 2 |
| TMQ | 2 | 2 |
| Calcium stearate | – | 3 |
| N326 carbon black | 30 | 99 |
| Diolpate 7017 | 6 | 34 |
| TBEP | – | 11 |

**Table 12.4** Sulphur, TBBS levels and compound properties from the designed experiment

| *Mix no.* | *Sulphur (phr)* | *TBBS (phr)* | *$M_H$–$M_L$ (dNm)* | *$t_{95}$ at 150 °C (min)* |
|---|---|---|---|---|
| 1 | 0.32 (−1) | 4.06 (+1) | 9.8 | 37.5 |
| 2 | 0.85 (0) | 1.50 (−1.414) | 10.0 | 10.0 |
| 3 | 0.85 (0) | 3.00 (0) | 12.2 | 13.0 |
| 4 | 1.38 (+1) | 4.06 (+1) | 15.8 | 12.0 |
| 5 | 0.32 (−1) | 1.94 (−1) | 7.9 | 21.3 |
| 6 | 0.85 (0) | 3.00 (0) | 12.5 | 12.5 |
| 7 | 0.85 (0) | 3.00 (0) | 14.1 | 13.0 |
| 8 | 0.85 (0) | 4.50 (+1.414) | 14.2 | 18.0 |
| 9 | 1.38 (+1) | 1.94 (−1) | 13.5 | 9.0 |
| 10 | 0.10 (−1.414) | 3.00 (0) | 5.8 | 48.5 |
| 11 | 0.85 (0) | 3.00 (0) | 12.7 | 13.3 |
| 12 | 1.60 (+1.414) | 3.00 (0) | 15.5 | 9.3 |

Figures in brackets refer to the variable level in design units

properties. The separately mixed masterbatches were cross-blended at 1:1 volume ratio and finalized in a single-stage in an internal mixer.

Vulcanizates were prepared by curing to rheometer $t_{max}$ at 150 °C and the resulting properties are shown in Table 12.5.

**Table 12.5** Properties of NR/Hv-BR/ENR-25 vulcanizates from the designed experiment

| *Mix no.* | *MR 100 (MPa)* | *T.S. (MPa)* | *E.B. (%)* | *Ring[a] fatigue (kc)* | *Cres. tear (N/mm)* | *Compression set (%)* | | *δ (°) 5% 5 Hz* | *Hard. (IRHD)* |
|---|---|---|---|---|---|---|---|---|---|
| | | | | | | *3d/23 °C* | *1d/70 °C* | | |
| 1 | 1.37 | 22.6 | 590 | 63 | 84 | 9 | 11 | 17.8 | 56 |
| 2 | 1.42 | 24.7 | 665 | 87 | 121 | 11 | 23 | 16.4 | 56 |
| 3 | 1.96 | 24.5 | 580 | 55 | 85 | 8 | 15 | 15.4 | 59 |
| 4 | 2.74 | 21.8 | 425 | 14 | 60 | 7 | 18 | 14.6 | 64 |
| 5 | 1.07 | 20.8 | 665 | 58 | 96 | 13 | 17 | 16.5 | 51 |
| 6 | 1.95 | 22.7 | 535 | 43 | 83 | 9 | 17 | 15.2 | 60 |
| 7 | 2.07 | 24.7 | 555 | 52 | 83 | 9 | 15 | 16.8 | 61 |
| 8 | 2.49 | 22.5 | 465 | 33 | 78 | 8 | 13 | 15.7 | 63 |
| 9 | 2.11 | 25.3 | 580 | 85 | 115 | 10 | 28 | 16.5 | 62 |
| 10 | 0.80 | 14.1 | 635 | 99 | 39 | 19 | 21 | 19.1 | 45 |
| 11 | 2.14 | 23.9 | 530 | 54 | 83 | 8 | 15 | 16.3 | 61 |
| 12 | 2.85 | 22.8 | 485 | 37 | 81 | 8 | 23 | 15.7 | 64 |

[a] 0–100% strain, 5 Hz, kc to failure

Response equations which relate the compound properties to design variables were generated using a commercial modelling program and are shown below:

Hardness, IRHD $= 60.2 + 5.7\ (S) + 2.1\ (TBBS) - 2.6\ (S)^2$

Tensile strength, MPa $= 24.2 + 2.1\ (S) - 2.4\ (S)^2$

Elongation at break, % $= 559.4 - 57.7\ (S) - 63.9\ (TBBS)$

Compression set, °
 1 day at 70 °C, % $= 16.2 + 2.6\ (S) - 3.8\ (TBBS) + 2.7\ (S)^2$

Cresent tear strength, N/mm $= 84.0 - 21.1\ (TBBS) - 15.4\ (S)(TBBS)$

Loss angle, 5%, 5 Hz, $= 16.3 - 1.0\ (S)$

Hardness increase at −26 °C
 after 28 days, IRHD $= 18.9 - 10.2\ (S) - 4.4\ (TBBS)$

Rate of hardness increase
 at −26 °C, IRHD per day $= 0.73 - 1.04\ (S) - 0.41\ (TBBS) + 0.62\ (S)^2$

where (S) and (TBBS) are the levels of curatives in design units.

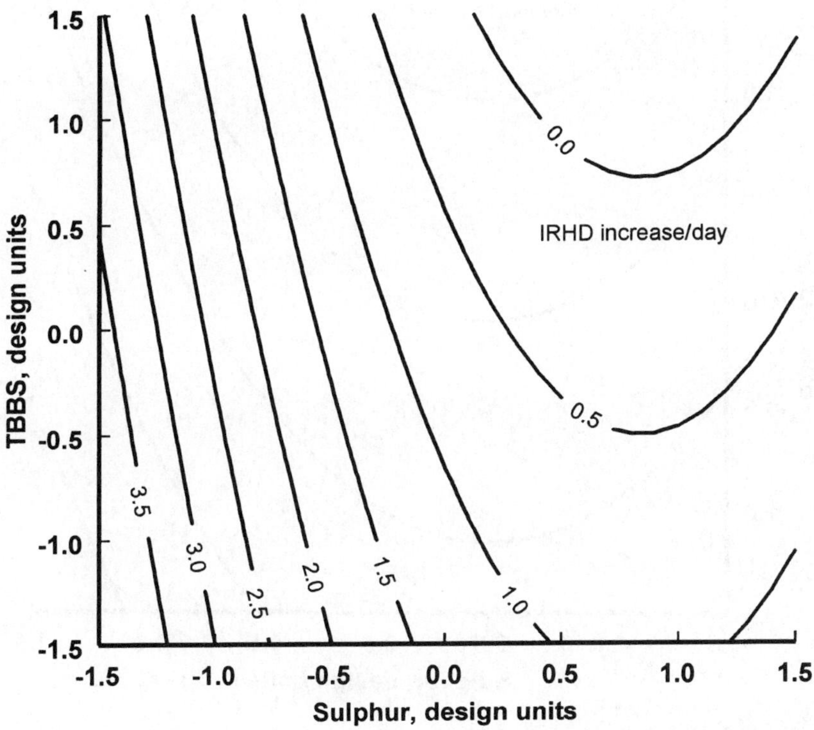


**Figure 12.4** Dependence of rate of hardness increase at −26 °C on sulphur and TBBS levels in NR/Hv-BR/ENR-25 vulcanizates.

The response equations in general indicate the expected trend in properties with crosslink level and type. For example, loss angle is reduced by increasing the sulphur level, i.e. total number of crosslinks. Crescent tear strength is more dependent on crosslink type; increasing the TBBS level, i.e. decreasing polysulphide crosslinks, decreases tear strength.

Contour plots may be generated from the response equations and show the response of properties to changes in sulphur and TBBS level. Figures 12.4 and 12.5 show the dependence of rate of hardness increase at −26 °C and compression set respectively, with sulphur and TBBS levels given in design units.

The figures serve to demonstrate the difficulty in achieving desirable values for two different properties as they indicate different design regions (S/TBBS levels) in which the most desirable properties are found, in this example resistance to low temperature crystallization and low compression set. The use of desirability functions in conjunction with response equations is the usual method of determining the most appropriate levels of variables to provide optimum properties [11]. For this

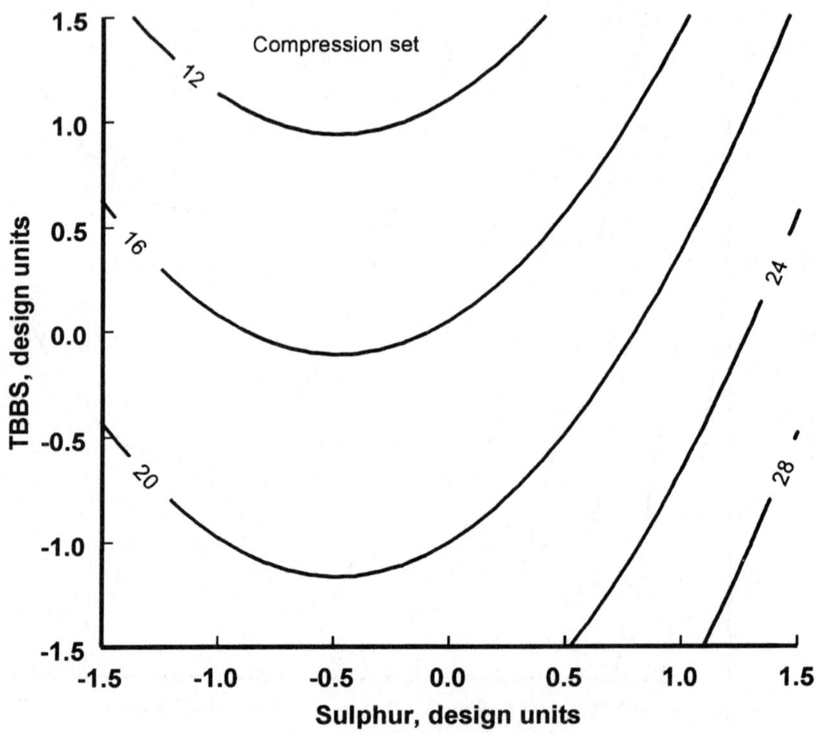


**Figure 12.5** Dependence of compression set (one day at 70 °C) on sulphur and TBBS levels in NR/Hv-BR/ENR-25 vulcanizates.

relatively simple two-variable system, superimposing contour plots allows a desirable region to be defined with a suitable degree of accuracy. This procedure was followed for all properties and the outcome indicated a S/TBBS level of 0.4/3.0 phr is the most appropriate for NR/Hv-BR/ENR-25 blends with high damping, low compression set and resistance to changes of properties at low temperature.

## 12.4 CONCLUSIONS

Both DSC and stress relaxation measurements demonstrate that NR/Hv-BR/ENR-25 blends have improved resistance to crystallization at low temperatures. Physical testing of bi- and tri-blends demonstrates that the addition of 10% Hv-BR does not severely affect the blend properties. A designed experiment to determine the optimum levels of curatives indicates that S and TBBS levels of 0.4 and 3.0 phr respectively are the most desirable in terms of overall vulcanizate properties.

## 12.5 REFERENCES

1 Brown, P.S. (1996) High damping NR/ENR blends. *Journal of Natural Rubber Research*, submitted for publication.
2 Crowther, B.G., Lewis, P.M. and Metherell, C. (1988) Chapter 6, Compounding, *Natural Rubber Science and Technology* (ed. A.D. Roberts). Oxford University Press, Oxford, p. 180.
3 Davies, C.K.L., Wolfe, S.V., Gelling, I.R. and Thomas, A.G. (1983) Strain crystallization in random copolymers produced by epoxidation of cis 1,4-polyisoprene. *Polymer*, **24**, 107–113.
4 Gelling, I.R. and Metherell, C. (1993) Vulcanisation systems for ENR-50. *Journal of Natural Rubber Research*, **8**(1), 37–56.
5 Gelling, I.R. and Morrison, N.J. (1985) Sulphur vulcanization and oxidative ageing of epoxidised natural rubber. *Rubber Chemistry and Technology*, **58**(2), 243–257.
6 Brown, P.S., Cook, S., Groves, S.A., Lewan, M.V. and Tinker, A.J. (1993) Natural rubber blends. Paper No. 127 presented at a meeting of the American Chemical Society, Rubber Division, Orlando, Florida, October 26–29.
7 Roland, C.M. (1988) Interactions in a miscible polymer blend. *Rubber Chemistry and Technology*, **61**(5), 866–878.
8 Stevenson, A. (1988) Chapter 18, Low temperature crystallization of natural rubber, in *Natural Rubber Science and Technology* (ed. A.D. Roberts). Oxford University Press, Oxford, 853–891.
9 Freakley, P.K. (1985) *Rubber Processing and Product Organisation*. Plenum Press, New York.
10 Davies, O.L. and Goldsmith, P.L. (1972) *Statistical Methods in Research and Production*. Oliver and Boyd, Edinburgh.
11 Derringer, G.C. (1983) A statistical methodology for designing elastomer formulations to meet performance specifications. *Kautschuk und Gummi Kunststoffe*, **35**(9), 349–352.

# 13

# Compounding NR/ENR-25 and NR-Hv-BR/ENR-25 blends for engineering applications

*Sally Groves*
*Tun Abdul Razak Research Centre, MRPRA, UK*

## 13.1 INTRODUCTION

The overall viability of blending normally compounded natural rubber (NR) with highly filled and plasticized 25 mole% epoxidized NR (ENR-25) to attain highly damping materials with a low dependence of properties on temperature is presented in Chapter 10. The aim of this chapter is to show how these novel blends may be processed successfully on a large scale and vulcanized satisfactorily in an injection-moulding process.

The general compounding requirements for these blend materials are short mix times, good carbon black dispersion and adequately small phase sizes. These criteria must be achieved when mixing on a large scale and it must be possible to tailor the properties of the blend to end-user requirements. It is essential that the two phases of the blend are prepared as separate masterbatches to ensure the required filler distribution. Masterbatches are then cross-blended and finalized with curatives. Initial work was undertaken in a small laboratory scale mixer (0.3 litres). It was at this scale that the use of designed experiments showed that N326 carbon black was the most desirable filler for both phases of these blends (Chapter 10). As the scale of mixing was increased, it became increasingly difficult to incorporate and adequately disperse the high levels of N326 black used in the ENR-25 phase within an acceptable time scale. No

*Blends of Natural Rubber*. Edited by Andrew J. Tinker and Kevin P. Jones.
Published in 1998 by Chapman & Hall, London. ISBN 0 412 819406

such problems were encountered with the lower levels of N326 black used in the NR masterbatch. Approaches to obtaining adequate dispersion in the ENR masterbatch and other refinements to the compound formulation are discussed here.

## 13.2 OPTIMIZING PLASTICIZER LEVEL AND MASTERBATCH RATIO

Work described in Chapter 10 revealed that plasticizer exudes from vulcanizates containing high levels of filler and plasticizer after several weeks. A combination of two plasticizers, Diolpate 7017 (Kemira Polymers, UK) and tri(butoxyethyl)phosphate (TBEP), has been shown to be preferable, but the former was identified as being responsible for bleeding. Therefore, it was necessary to determine partition coefficients ($K_d$) for Diolpate 7017 between black-filled NR and ENR-25 vulcanizates as described in Chapter 11. The $K_d$ values obtained indicated that the maximum plasticizer level to be added to each of the polymer masterbatches is lower than expected from $K_d$ values found previously in work on gum vulcanizates [1].

It is important to know the effect of reducing plasticizer levels on both processing behaviour and properties of the blends. This was established in a two-variable designed experiment in which the volume ratio of polymer masterbatches used in the blend was the second variable. Some interaction between these two variables is expected when bleeding of plasticizer is considered. The methodology of designed experiments is explained in Chapter 10 and elsewhere [2, 3].

The basic formulation of masterbatches used is given in Table 13.1. The plasticizer level in the ENR-25 masterbatch and volume ratios of NR and ENR-25 masterbatches used in the blends are given in Table 13.2.

Compounds were prepared by the procedures described in Chapter 10 and vulcanizates were cured to rheometer $t_{95}$ at 150 °C. The physical properties obtained for the vulcanizates are shown in Table 13.3.

The results were analysed using regression analysis, and of the properties measured only three produced valid response equations: these are shown below.

$$\text{Elongation at break, \%} = 480 + 24x$$
$$\text{Loss angle, 5\% strain, 5Hz, }^\circ = 18.55 - 1.17x$$
$$\text{Tensile strength, MPa} = 19.77 + 1.64x$$

where $x$ is the percentage of the NR masterbatch in the blend by volume expressed in design units (Table 13.2).

No response terms with valid $y$ coefficients were determined. Therefore reducing plasticizer level within the modest range considered has little effect on physical properties or damping as given by the loss angle, $\delta$.

**Table 13.1** NR and ENR-25 masterbatch formulations

| | | |
|---|---|---|
| NR | 100 | – |
| ENR-25[a] | – | 100 |
| N326 carbon black | 30 | 99 |
| Diolpate 7017[b] | 6 | variable |
| TBEP | – | 11 |
| Zinc oxide | 5 | 5 |
| Stearic acid | 2 | 2 |
| TMQ | 2 | 2 |
| Calcium stearate | – | 3 |
| Crossblended at a 1:1 volume ratio | | |
| Sulphur | | 0.3 |
| TMTD | | 2.0 |
| TBBS | | 3.0 |

[a] Epoxyprene-25, Kumpulan Guthrie Berhad, Malaysia
[b] A polymeric ester, Kemira Polymers, Stockport, Cheshire, UK

**Table 13.2** Plasticizer level in ENR-25 masterbatch and masterbatch volumes used in the designed experiment

| *Mix no.* | *Design units* $x$ | $y$ | *NR masterbatch (% by volume), x* | *Diolpate 7017 (phr), y* | *Bleeding observed* |
|---|---|---|---|---|---|
| D1 | 1 | 1 | 60 | 35 | ✓after 3wks |
| D2 | 1 | –1 | 60 | 29 | × |
| D3 | –1 | 1 | 40 | 35 | ✓ |
| D4 | –1 | –1 | 40 | 29 | ✓ |
| D5 | 0 | 1.414 | 50 | 36.2 | ✓ |
| D6 | 0 | –1.414 | 50 | 27.8 | × |
| D7 | 2.4 | 0 | 74 | 32 | × |
| D8 | –2.4 | 0 | 26 | 32 | ✓ |
| D9 | 0 | 0 | 50 | 32 | slightly sticky |
| D10 | 0 | 0 | 50 | 32 | × |
| D11 | 0 | 0 | 50 | 32 | × |
| D12 | 0 | 0 | 50 | 32 | × |

Processing with lower plasticizer levels proceeded smoothly, and thus the level of Diopate plasticizer used may be reduced slightly. However, the effects of reducing plasticizer level on dynamic properties at low

**Table 13.3** Properties of NR/ENR-25 vulcanizates from the designed experiment

| *Mix no.* | *Hard (IRHD)* | *MR100 (MPa)* | *T.S (MPa)* | *E.B. (%)* | *Compression set (%)* | | *Cres. tear* | *δ(°) 5%* |
|---|---|---|---|---|---|---|---|---|
| | | | | | *3 d/23 °C* | *1 d/70 °C* | *(N/mm)* | *5 Hz* |
| D1 | 59 | 2.05 | 21.5 | 490 | 13 | 23 | 83 | 17.5 |
| D2 | 61 | 1.89 | 21.7 | 525 | 13 | 22 | 70 | 17.1 |
| D3 | 60 | 1.84 | 18.4 | 480 | 13 | 26 | 79 | 19.4 |
| D4 | 62 | 2.19 | 18.5 | 460 | 12 | 21 | 87 | 20.2 |
| D5 | 60 | 2.02 | 20.0 | 475 | 14 | 23 | 77 | 18.7 |
| D6 | 62 | 2.07 | 18.6 | 460 | 13 | 24 | 77 | 18.6 |
| D7 | 58 | 1.93 | 23.5 | 520 | 11 | 19 | 69 | 15.6 |
| D8 | 63 | 2.11 | 15.5 | 395 | 14 | 24 | 85 | 20.8 |
| D9 | 59 | 2.00 | 20.8 | 500 | 12 | 25 | 82 | 19.0 |
| D10 | 60 | 2.01 | 19.5 | 480 | 12 | 23 | 72 | 19.0 |
| D11 | 61 | 1.94 | 19.6 | 500 | 12 | 22 | 79 | 18.7 |
| D12 | 62 | 2.08 | 19.6 | 485 | 11 | 24 | 83 | 18.0 |

temperatures may be more deleterious due to an effective increase in $T_g$ of the ENR-25 phase.

Masterbatch ratio was a significant factor in determining the damping of the blend vulcanizate. To increase damping the level of NR should be reduced; however, a compromise must be made because bleeding is seen in blends with high percentages of ENR-25 masterbatch (Table 13.1). The best properties with no bleeding were obtained at a 1:1 by volume ratio of masterbatches and 32 phr of Diolpate in the ENR-25 masterbatch.

## 13.3 MIXING TRIALS

It is essential that carbon black masterbatches of the two polymers are mixed separately to ensure the required distribution of filler. The masterbatches are then cross-blended and finalized with curatives in two further mixing stages in an internal mixer. The initial work in a Brabender Plasticorder, with a gross capacity of about 0.4 litre, showed that N326 is the most desirable grade of carbon black for both the NR and ENR-25 phases (Chapter 10). The scale of mixing was increased to a 00C Banbury, with a gross capacity of about four litres, with little need for change to the processing conditions, although the dispersion of carbon black was poorer for the ENR-25 masterbatch. As scale was increased further to a K2A Intermix with intermeshing rotors and a gross capacity of about 40 litres, the only practicable way of incorporating the large volumes of plasticizer necessary was by injecting directly into the mixing chamber. In practice this necessitated the use of identical plasticizer

**Table 13.4** Revised masterbatch formulations[a]

| | | |
|---|---|---|
| SMR CV60 | 100 | – |
| ENR-25 | – | 100 |
| Calcium stearate | – | 5 |
| N231 | 30 | 100 |
| Diolpate 7017 | 3 | 30 |
| TBEP | 1 | 10 |

[a] Both masterbatches also contain 3.5 phr zinc oxide, 1.5 phr stearic acid, 2.5 phr TMQ, 3 phr IPPD and 4 phr Sunproof Improved wax

systems in each of the NR and ENR-25 masterbatches. The revised masterbatch formulations are shown in Table 13.4.

NR and ENR-25 masterbatches were then cross-blended at 1:1 volume ratio in an internal mixer and finalized with curatives in a separate stage again in an internal mixer. Optimum levels of curatives to achieve the highest damping for the lowest compression set were established in work described in Chapter 12 and are 0.4 phr S and 3 phr TBBS.

Another problem encountered when mixing at large scales was the need to use grades of carbon black with higher structures than N326 in order to achieve good black dispersion in the ENR-25 masterbatch. N231 black was used initially because the designed experiment described in Chapter 10 indicated that it would give comparable properties to those obtained with N326. However N231 is currently only available in the United States and for this reason N330, the next preferred alternative identified by the designed experiment, was also used. The properties of blends prepared with equivalent levels of the three different grades of carbon black are presented in Table 13.5.

**Table 13.5** Effect of grade of carbon black used in the ENR-25 masterbatch on blend properties

| *Grade* | *N326* | *N231* | *N330* |
|---|---|---|---|
| *Black dispersion, Cabot rating* | *C1–3* | *B1–3* | *B1-3* |
| Hardness, IRHD | 56 | 52 | 59 |
| Tensile strength, MPa | 22 | 24 | 22 |
| Elongation at break, % | 660 | 660 | 630 |
| Crescent tear strength, N/mm | 100 | 110 | 100 |
| Compression set, 1 d/70 °C, % | 15 | 18 | 19 |
| Loss angle, 5 Hz, 5% strain, ° | 20 | 20 | 18 |

**Table 13.6** Masterbatch mix cycles used in a Farrel F50 Banbury

| | *NR* | *ENR-25* |
|---|---|---|
| Rotor speed | 45 rpm | 45 rpm |
| Start | polymer | polymer, powders, black |
| 1.5 min | powders, black | plasticizer |
| 2 min | plasticizer | |
| | 60 rpm | 60 rpm |
| | 155 °C dump | 150 °C dump |
| Mix time | 3.5 minutes | 4 minutes |
| Mooney viscosity, ML(1+4), 100 °C | 55 | 45 |

**Table 13.7** Crossblending/finalizing cycles in a Farrel F50 Banbury

| | *Cross-blending* | *Finalizing* |
|---|---|---|
| Rotor speed | 50 rpm | 30 rpm |
| Start | masterbatch | crossblend, curatives |
| 1 min | ram lift | ram lift |
| | dump | dump |
| Mix time | 2 minutes | 2 minutes |
| Dump temperature | 140 °C | 90 °C |
| Mooney viscosity, ML(1+4), 100 °C | 42 | 30 |

N326 and N231 give similar vulcanizate properties, but compound containing N231 has a better dispersion rating. N330 gives a significantly reduced loss angle, around 2° lower than the other grades of carbon black, but dispersion is good. The recommendation for these materials is to use N231 wherever possible. Further development work on larger mixers may determine processing conditions to enable successful incorporation of N326 filler in the future. Further work on this is advised.

Pilot scale mixing has also been performed in a 50 litre Farrel F50 Banbury fitted with Synchronous Technology (ST) rotors. The mixing conditions used are given in Tables 13.6 and 13.7. The properties of the vulcanizates derived from this processing are shown in Table 13.8 under dual stage processing. It has been possible to achieve both the desired high damping combined with low compression set and good general physical properties from compounds mixed on a large scale.

**Table 13.8** Properties of vulcanizates prepared from compounds mixed in a Farrel F50 Banbury

| *Crossblending and finalizing process* | *Dual stages* | *Single stage* |
|---|---|---|
| Dispersion, Cabot | A1 | A2 |
| Hardness, IRHD | 51 | 54 |
| MR100, MPa | 1.3 | 1.5 |
| Tensile strength, MPa | 21.0 | 22.5 |
| Elongation at break, % | 690 | 680 |
| Compression set, 1d/70 °C, % | 17 | 17 |
| Crescent tear, N/mm | 75 | 85 |
| Ring fatigue, 0–100%, kc to failure | 140 | 120 |
| Loss angle, 5 Hz, 5% strain, ° | 19 | 18 |
| G*, 5 Hz, 5% strain, MPa | 2.5 | 2.6 |
| G*(–25 °C)/G*(30 °C), 5 Hz, 5% strain | 3.0 | 2.5 |

## 13.4 SINGLE-STAGE MIXING FOR CROSS-BLENDING AND FINALIZING

The mixing schedules adopted thus far involve a rather lengthy process to achieve a final compound. To assess whether less well mixed materials would have suitable properties, mixing was performed in the Farrel F50 Banbury mixer by two different processes. The first was the usual separate cross-blending and finalizing stages (Tables 13.6 and 13.7), whilst the cross-blending and finalizing with curatives were combined in one stage in the second. The conditions used for this single-stage process are shown in Table 13.9.

The properties of vulcanizates prepared by the single-stage cross-blend/finalize procedure are included in Table 13.8. Comparing these with the properties of comparable vulcanizates prepared from compound cross-blended and finalized in two stages indicates that the two materials are very similar. The slightly higher compound viscosity,

**Table 13.9** Mixing cycle for single stage crossblending/finalizing in a Farrel F50 Banbury mixer

| | |
|---|---|
| Rotor speed | 30 rpm |
| Start | polymers, curatives<br>ram lift |
| Mix time | 2 minutes |
| Dump temp | 95 °C |
| Mooney viscosity, ML(1+4), 100 °C | 35 |

hardness and stiffness values for the one-stage material are the to-be-expected consequences of less mixing [4, 5], and the slightly lower damping may be ascribed to less breakdown of the rubber in the compound. The slightly lower fatigue life may be due to a combination of the slightly higher modulus increasing the severity of the constant strain fatigue test and the difference in black dispersion. Although black dispersion is slightly poorer for the compound cross-blended and finalized in a single stage, it is still well within the expectations of factory mixing. These observations suggest that a one-stage cross-blending/finalizing procedure may be effective in full-scale mixing of these compounds.

## 13.5 NR/Hv-BR/ENR-25 BLENDS

In order to reduce the low temperature crystallization of the NR phase in NR/ENR-25 blends, high vinyl butadiene rubber (Hv-BR) has been used to replace 10% by weight of the NR when preparing this black masterbatch (Chapter 12). The Hv-BR is miscible with NR, hence vulcanizates showed similar properties to comparable NR/ENR-25 blends and an improvement in resistance to low temperature crystallization. In this earlier work, VI 1979 (Bayer, 80% 1,2-content) was used, however production of this grade ceased and a different grade, VI 1969 (Bayer, 70 ± 3% vinyl), was substituted in work on larger-scale mixing. The properties of a tri-blend containing VI 1969 are compared with those of a similar NR/ENR-25 blend in Table 13.10. The comparison confirms that the close similarity in physical properties and improvement in resistance to low temperature crystallization found in the laboratory are also obtained when compound is mixed on a large scale.

**Table 13.10** Comparison of properties of NR/ENR-25 and NR/Hv-BR/ENR-25 blends prepared on a 40 litre scale

| *Property* | *NR/ENR-25* | *NR/Hv-BR/ENR-25* |
|---|---|---|
| Hardness, IRHD | 57 | 56 |
| Tensile strength, MPa | 23.5 | 22.5 |
| Crescent tear strength, N/mm | 100 | 100 |
| Compression set, 1d/70 °C, % | 17 | 15 |
| Loss angle, 5 Hz, 5% strain, ° | 19 | 19 |
| Increase in IRHD, −26 °C: | | |
| 1 day | 15 | 9 |
| 4 days | 25 | 21 |
| 7 days | 27 | 23 |

## 13.6 INJECTION MOULDING

The components for which these blend materials are being developed are generally produced by injection-moulding. This process requires a combination of adequate scorch safety and high cure rates at operating temperatures of 170–180 °C in order to achieve the necessary short moulding cycle times. The higher cure temperatures will affect vulcanizate properties, even when an EV cure system is used. The suitability of these blends for injection-moulding was assessed initially using compound formulated according to Table 13.4, but with 10 phr of Hv-BR in the NR masterbatch and 0.4 phr S/3 phr TBBS as cure system. Injection moulding was carried out in a REP B43K injection-moulding machine using the settings in Table 13.11. Tensile properties were determined on test pieces cut tangentially to the circular flow pattern in a sheet moulded in a centre-gated cavity. Hardness buttons and compression set pellets came from a multi-cavity mould. It was necessary to produce double shear test pieces by transfer moulding at 180 °C. Vulcanizate properties are summarized in Table 13.12.

**Table 13.11** Injection mould machine settings

| *Setting* | *Value* |
|---|---|
| Screw temperature | 80 °C |
| Ram chamber temperature | 90 °C |
| Nozzle block temperature | 60 °C |
| Nozzle diameter | 3 mm |
| Injection time | 3–4 secs |
| Mould temperature | 180 °C |

**Table 13.12** Properties of NR/Hv-BR/ENR-25 blend vulcanizate

| *Moulding condition* | *Press at 150 °C* | *Injection at 180 °C* |
|---|---|---|
| Hardness, IRHD | 56 | 54 |
| M100, MPa | 1.35 | 1.10 |
| Tensile strength, MPa | 22 | 22 |
| Elongation at break, % | 640 | 680 |
| Compression set: | | |
| 3 days at 23 °C, % | 18 | 32 |
| 1 day at 70 °C, % | 16 | 22 |
| Loss angle[a], 5% strain, 5 Hz, 23 °C, ° | 19 | 21 |

[a] Double shear test pieces, transfer moulded at the given moulding temperature

**Table 13.13** Unaged properties of injection moulded NR/Hv-BR/ENR-25 blend vulcanized with a cure system adjusted for high temperatures

| *Property* | |
|---|---|
| Hardness, IRHD | 53 |
| M100, MPa | 1.2 |
| Tensile strength, MPa | 18 |
| Elongation at break, % | 590 |
| Compression set: | |
| 3 days at 23 °C, % | 13 |
| 1 day at 70 °C, % | 17 |
| 1 day at 100 °C, % | 27 |
| 1 day at 120 °C, % | 44 |
| $G_{stat}$, 0–50%, MPa | 1.2 |
| G*, 5% strain, 5 Hz, 23 °C, MPa | 1.5 |
| Loss angle, 5% strain, 5 Hz, 23 °C, ° | 18 |
| Tan(δ), 5% strain, 5 Hz, 23 °C | 0.32 |

Comparison with the properties of a similar compound cured at 150 °C, which are included in Table 13.12, confirms the expected effects of curing at a higher temperature: reduced moduli, increased compression set and increased damping. All are due to a lower overall crosslink density. An increase in curative levels and the use of secondary accelerators to reduce reversion was necessary to redress these changes. The same basic formulation was employed, but with a cure system consisting of 0.33 phr S, 1.1 phr TBBS and 3.3 phr N-*t*-butyl-2-benzothiazyl sulphenimide (TBSI). The properties of injection-moulded vulcanizate are presented in Table 13.13. The recovery properties are at least comparable to those of the previous compound vulcanized at 150 °C (Table 13.12) and the loss angle is identical. The low dynamic:static modulus ratio evident from data in Table 13.13 is noteworthy; the ratio of 1.25 may be compared directly with the data presented in Figure 10.5 of Chapter 10.

## 13.7 CONCLUSIONS

A designed experiment has identified both appropriate levels of plasticizer and the optimum NR(Hv-BR):ENR-25 masterbatch ratio for the desired balance of physical properties and lack of bleeding from vulcanizate. A 1:1 volume ratio is best, and a reduction in Diolpate 7017 plasticizer level to 32 phr in the ENR-25 masterbatch ensures no bleeding at this masterbatch ratio and no significant adverse effect on either properties or processing behaviour.

These highly filled and plasticized blends have been mixed successfully in mixers up to 50 litres in gross capacity. The only change seen to be necessary is a move from N326 carbon black to either N231 or, less desirably, N330 in order to obviate difficulties in dispersing the high loadings of carbon black in the ENR-25 masterbatch. Replacement of N326 with N231 has little adverse effect on vulcanizate properties, but the use of N330 increases hardness and reduces damping slightly. Tri-blends incorporating 10% Hv-BR in the NR masterbatch may also be mixed on a large scale without difficulty or loss of efficacy in the improvement of resistance to low temperature crystallization.

If the good recovery properties obtained in press curing at 150 °C are to be retained in injection-moulding at temperatures in the region of 180 °C, it is necessary to increase overall curative levels and substitute a more reversion resistant accelerator for a major part of the TBBS used previously. If compression set is not a major consideration, higher damping may be obtained by continuing to use the cure system optimized for press curing at 150 °C.

In summary, good practical formulations have been developed; they are suitable for large-scale mixing, process well by both press curing and injection-moulding, and provide the desired combination of high damping, good physical properties, low set and low dynamic:static modulus ratio.

## 13.8 REFERENCES

1 Aris bin Ahmad, Tinker, A.J. and Aubrey, D.W. (1995) Distribution of ester plasticisers between two rubber phases – estimation of partition coefficients and comparison with theory. *Journal of Natural Rubber Research*, **10**(1), 1–13.

2 Freakley, P.K. (1985) *Rubber Processing and Product Organisation*. Plenum Press, New York.

3 Davies, O.L. and Goldsmith, P.L. (1972) *Statistical Methods in Research and Production*. Oliver and Boyd, Edinburgh.

4 Boonstra, B.B. and Medalia, A.I. (1963) Effect of carbon black dispersion on the mechanical properties of rubber vulcanizates. *Rubber Chemistry and Technology*, **36**(1), 115–142.

5 Aris bin Ahmad, Tinker, A.J. and Bristow, G.M. (1989) Influence of compound properties on vulcanisate properties of black-filled natural rubber, in *Proceedings, International Rubber Technology Conference 1988 Penang, Malaysia* (eds L.L. Amin and K.T. Lau), Rubber Research Institute of Malaysia, Malaysia.

# 14

# Solutions to the basic problems of poor physical properties of NR/EPDM blends

*Stuart Cook*
*Tun Abdul Razak Research Centre, MRPRA, UK*

## 14.1 INTRODUCTION

The development of blends of natural rubber (NR) with ethylene-propylene-diene monomer (EPDM) with the aim of combining the excellent physical properties of NR with the ozone resistance of EPDM has received much attention over the past three decades [1–7]. The principal target application for such blends has for many years been tyre sidewalls [8] where ozone cracking was seen as a limiting factor on the service life of a tyre. However, extruded weatherseal profiles for vehicles and light-coloured injection-moulded goods for domestic appliances, markets once dominated by NR, but now almost entirely superseded by EPDM, are also attractive areas for NR/EPDM blends. Whilst ozone resistance has been comparatively easy to achieve in NR/EPDM blends [9], the general physical properties of NR/EPDM blends have been less than satisfactory [10]. This is entirely expected as it is generally the case that the physical properties of vulcanized elastomer blends are inferior to those predicted from the properties of the component elastomers [11–13].

The main problems associated with the generally poor properties of NR/EPDM blends are a direct result of the relative chemical reactivities of NR and EPDM. NR is a highly unsaturated and hence reactive polymer, making it very susceptible to attack by ozone; conversely EPDM has

*Blends of Natural Rubber*. Edited by Andrew J. Tinker and Kevin P. Jones
Published in 1998 by Chapman & Hall, London. ISBN 0 412 819406

a very low level of unsaturation, typically less than 3 mole%, and hence has a low reactivity towards ozone. In a blend, such a disparity in reactivity is reflected in the relative rates of crosslinking during vulcanization [6, 13–15], both within and between the two phases, diffusion of curatives towards the faster curing NR phase occurring readily during mixing and curing [16]. The lower solubility of many curatives in EPDM [5, 16] as compared with NR further reduces the propensity towards crosslink formation in the EPDM phase. Overall, these factors combine to give a blend having a highly crosslinked NR phase and a very poorly crosslinked EPDM phase [17, 18].

## 14.2 APPROACHES TO IMPROVING NR/EPDM BLEND PROPERTIES

There have been several notable approaches to improving the properties of NR/EPDM blends attempted over the past 25 years or so. Generally, these approaches have sought to increase the cure rate of EPDM, either by means of modifying the EPDM to make it more reactive towards curatives, or by using curatives that have an increased reactivity towards EPDM. Morrissey [1] reported the halogenation of EPDM in solution to be effective in improving cure compatability in blends with highly unsaturated elastomers such as NR and SBR. Baranwal and Son [2] modified EPDM in solution with accelerator species, including the sulphur donor dithiodimorpholine (DTDM), using UV irradiation in the presence of a photosensitizer. Similar work was reported by Hashimoto *et al.* [3] whereby pendant accelerator groups were introduced to the EPDM polymer chain prior to cross-blending with NR. Hopper [4] modified EPDM with N-chlorothioamides, again in solution, to introduce reactive pendant groups which could behave as a crosslink precursor during normal vulcanization of the blend. More recently, Coran [7] has reported the modification of EPDM with maleic anhydride in the internal mixer at high temperatures which, through the introduction of carboxylate pendant groups, was suggested to form reversible ionomeric crosslinks with zinc oxide added as part of the normal sulphur cure system.

The increasing commercial availability of EPDM grades with higher levels of unsaturation, aimed specifically at increasing the rate of cure in EPDM compounds with sulphur-based cure systems, has also led to improved levels of cure being achieved in the EPDM phase of a blend with NR. However, even the so-called 'very highly' unsaturated EPDM grades still have less than 10% of the unsaturation of NR on a molar basis.

As an alternative to modifying EPDM to redress the imbalance in reactivity of curatives towards NR, the use of accelerators with greater solubility in the EPDM phase of a blend has been investigated [5, 6]. Although the use of accelerators containing higher alkyl substitution was shown to greatly enhance the properties of blends containing EPDM, few if any of these accelerators with long alkyl substituents have been commercialized. The mobility of curatives between phases of a blend of elastomers which differ in polarity can be restricted by the use of lead oxide activator [14]. The resulting insoluble lead accelerator salt effectively removes the thermodynamic driving force for curative diffusion between the elastomer phases and gives a blend with a good level of covulcanization. Manipulation of curative type, whilst not requiring an additional mixing stage, does however preclude the means for imparting other useful properties on the blend that EPDM modification can produce [1].

All of the EPDM modification methods described above achieve a significant improvement in the physical properties of NR/EPDM blends. This results directly from the introduction of reactive sites able to participate in the cure process and speed up vulcanization in the EPDM phase, so improving the level of crosslinking achieved. It should be noted, however, that only one of the EPDM modification procedures recorded above was achieved in an internal mixer, although the reagents and conditions required for modification are hardly suitable for large-scale mixing in a factory environment.

The general approach adopted in the work described in this chapter was one developed from the work of Baranwal and Son [2] and more especially Hopper [4]. Three commercially available sulphur donors, *bis*-alkylphenoldisulphide (BAPD), dithiodicaprolactam (DTDC) and dithiodimorpholine (DTDM), have been used to modify EPDM by mixing at elevated temperatures in an internal mixer as part of a normal masterbatch mixing cycle. This procedure has been termed 'reactive mixing'. As has been the case in work reported by others [2, 4, 7] in this field, the choice of EPDM grades used has been restricted to commercially available grades with a very high level of unsaturation. The modification is believed to be similar in nature to that described by Hopper [4] in which a functionality (alkylphenol monosulphide, caprolactam or morpholino) derived from the sulphur donor is bound to the EPDM via a sulphur linkage. A simplified representation of what is thought to occur is shown in Figure 14.1a. During vulcanization, either part, or all, of the bound sulphur donor moiety can act as a leaving group, being displaced by 2-mercaptobenzothiazole (MBT), a vulcanization intermediate, to form a crosslink precursor site on the EPDM polymer chain (Figure 14.1b).

For all of the work described in Chapters 15 to 17, dealing with applications of NR/EPDM blends, a 60:40 NR:EPDM blend ratio was used. This blend ratio has been shown to exhibit good ozone resistance in both static and dynamic ozone tests. It is desirable both in terms of performance and cost to maximize the NR content of such blends whilst maintaining adequate ozone protection without added antiozonants. Whilst lower volume ratios of EPDM have been investigated, indications

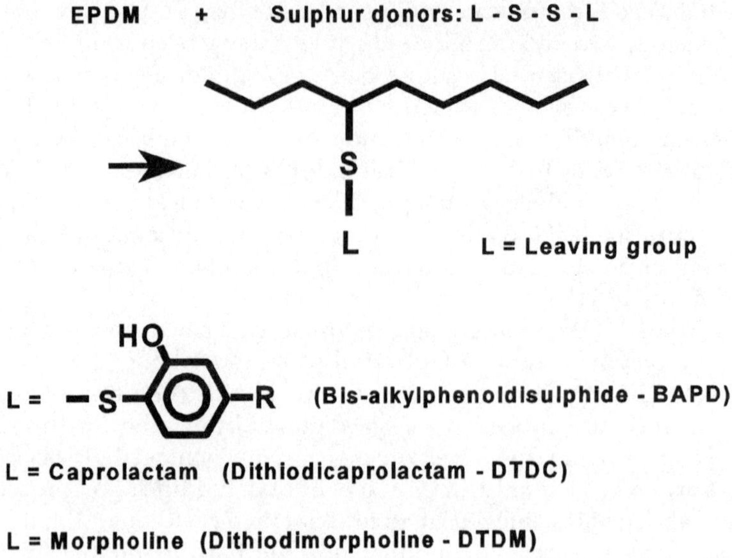


**Figure 14.1a** General approach to EPDM modification.

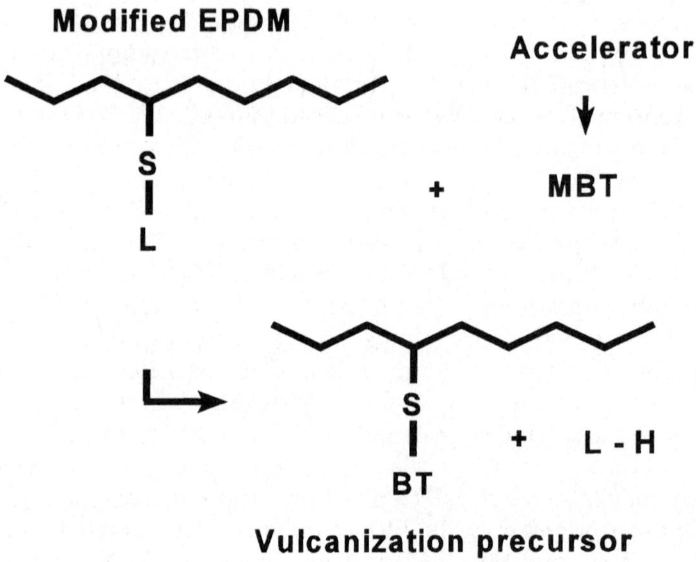


**Figure 14.1b** Formation of vulcanization precursor.

were that insufficient ozone protection was afforded. These findings are in accord with previously published work [19]. The work described here demonstrates that modification of EPDM with sulphur donors using a 'reactive mixing' procedure leads to an improved level of crosslinking in the EPDM phase of an NR/EPDM blend. For one of the modifiers (BAPD), a reduction in phase size in the blend is also observed. In addition, all three of the modifiers act so as to increase the level of interaction between the EPDM and carbon black.

## 14.3 MIXING PROCEDURES

For the compounds described in this chapter, mixing was performed in a variety of laboratory scale internal mixers ranging in size from a Brabender PL2000E fitted with a 350S mixing head and Banbury style rotors (300 ml capacity, 0.75 fill factor) to a 00C Banbury (3 litre capacity, 0.75 fill factor). In order for modification of EPDM to be achieved, separate NR and EPDM masterbatch mixing procedures have been adopted. Cross-blending of the modified EPDM masterbatch with a NR masterbatch is achieved in a second mixing stage and a third mixing stage is required for the addition of curatives. The EPDM modification procedure is carried out using $\sim$1 phr modifier in an internal mixer under conditions which will result in a batch temperature of at least 150 °C being achieved and then maintained for between 1 and 2 minutes. The batch temperature on dumping is typically $170° \pm 5°$ which is not unusual for factory processing.

The requirements of the procedure are consistent with the modification reaction being free radical in nature, in that high temperatures facilitate cleavage of the sulphur donor S—S bond and the formation of polymer radical species. The admittance of air at such a stage of the mixing procedure has been found to effectively quench the modification reaction, oxygen being an extremely efficient radical trap. Nonetheless, the necessary conditions can be attained in a factory environment – as demonstrated by mixing in a No. 11 Banbury, as described in Chapter 15.

Analysis of the modified EPDM has been carried out by gel permeation chromatography (GPC) followed by ultraviolet (UV) spectroscopy at 240 nm. This has confirmed that UV active groups have become bound to the EPDM polymer chain. Quantification of the bound modifier is possible if the extinction coefficient of the bound moiety is taken to be that of the sulphur donor. Typically the level of modification has been estimated by the GPC/UV technique to be in the range 0.1–0.5 wt%. However, at such a low level of modification, conclusive structural identification of the polymer bound groups has not been possible.

## 14.4 EFFECT OF EPDM MODIFICATION ON CROSSLINK DENSITY

Determination of the level of crosslinking in the EPDM phase of a blend with NR has been achieved by using the technique of swollen-state $^{13}C$ nuclear magnetic resonance spectroscopy (NMR) [17], as outlined in

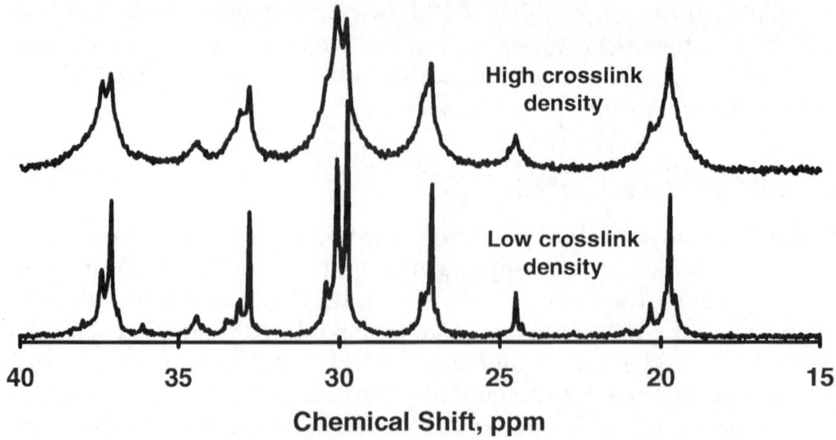


**Figure 14.2** $^{13}C$ NMR spectrum of EPDM gum vulcanizates.

Chapter 2. Essentially, the effect of increasing the crosslink density of EPDM is to increase peak widths in the $^{13}C$ NMR spectrum of the swollen vulcanizate. This can be seen in Figure 14.2, which shows the aliphatic region of the $^{13}C$ NMR spectra of two EPDM vulcanizates containing different curative levels. Measurements of spectral peak widths can be quantified and related to physical crosslink densities as measured by classical stress–strain measurements [20]. The relationship between peak width, as defined by the peak height:valley height ratio at 29.78 ppm, and physical crosslink density for EPDM can be seen in Figure 14.3. A calibration curve obtained in this way can then be used to

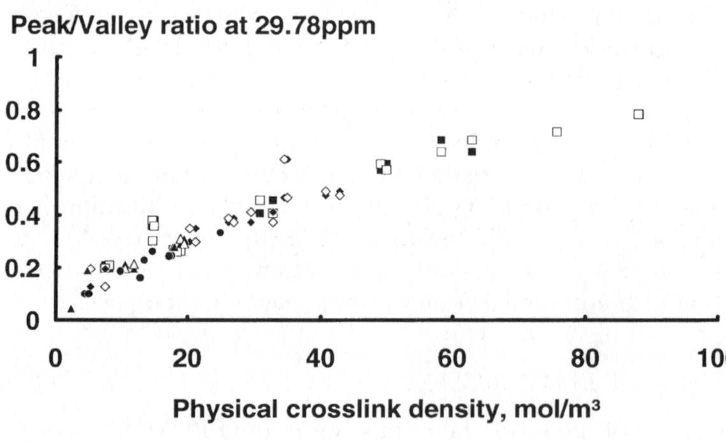


**Figure 14.3** $^{13}C$ NMR calibration curve for different 'highly unsaturated' EPDM grades.

determine the physical crosslink density of EPDM in a blend where physical crosslink densities cannot be obtained reliably from stress–strain measurements. Several different EPDM grades, all having similar very high levels of unsaturation, have been analysed and found to lie on the same calibration curve in the region between 5 and 35 mol/m$^3$.

The effect of EPDM modification on the crosslink density in the EPDM phase of a 60:40 NR:EPDM gum (unfilled) blend vulcanizate can be correlated with the tensile strength of the blend. For vulcanizates cured to $t_{max}$ at 150 °C by using a conventional 2.0 phr sulphur/0.6 phr CBS cure system, it can be seen in Figure 14.4 that modification by each one of the sulphur donors results in a significant increase in both crosslink density in the EPDM phase and tensile strength. Typically, the effect of EPDM modification is to increase the level of crosslinking in the EPDM phase from about 12 mol/m$^3$ to about 22 mol/m$^3$. Whilst this is still a long way short of the level of crosslinking in the NR phase (typically 80 mol/m$^3$), the increase in crosslink density seen in the EPDM phase is sufficient to bring about considerable improvements in tensile strength. It is of note that the tensile strength and EPDM crosslink density of gum blends prepared by 'reactive mixing' are very similar [17] to those of blends prepared by Hopper [4] and by the methods of Coran [7]. It would thus appear that a 'threshold' crosslink density of about 20 mol/m$^3$ needs to be achieved in the EPDM phase before any significant improvement in physical properties of the blend is noted [17]. Whilst black-filled blends have not been analysed quantitatively by $^{13}$C NMR, similar indications are seen.

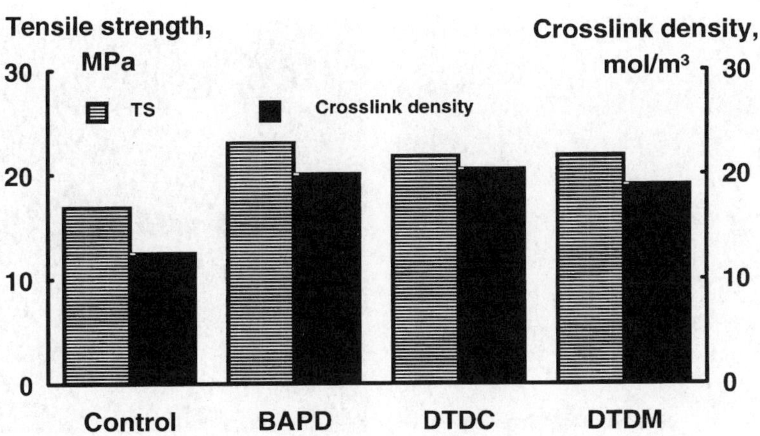


**Figure 14.4** Effect of EPDM modification on gum blend properties (2.0 phr S/0.6 phr CBS, $t_{max}$/150 °C).

## 14.5 EFFECT OF EPDM MODIFICATION ON PHASE MORPHOLOGY

Figure 14.5 shows STEM micrographs (see Chapter 3) of 60:40 NR:EPDM gum blend vulcanizates. It can clearly be seen that modification by BAPD (b) leads to a significantly smaller blend phase size than in the unmodified EPDM blend (a). Whilst accepting that the measurement of phase size is somewhat imprecise, it is estimated that the BAPD modification of EPDM reduces the typical phase size in the blend from about 3 μm to about 1 μm. Modification of EPDM by either DTDC (c) or DTDM (d) appears to have less of an effect upon phase size, although there are differences apparent in the phase morphology between the modified blends and the unmodified control blend. Black-filled blends are believed

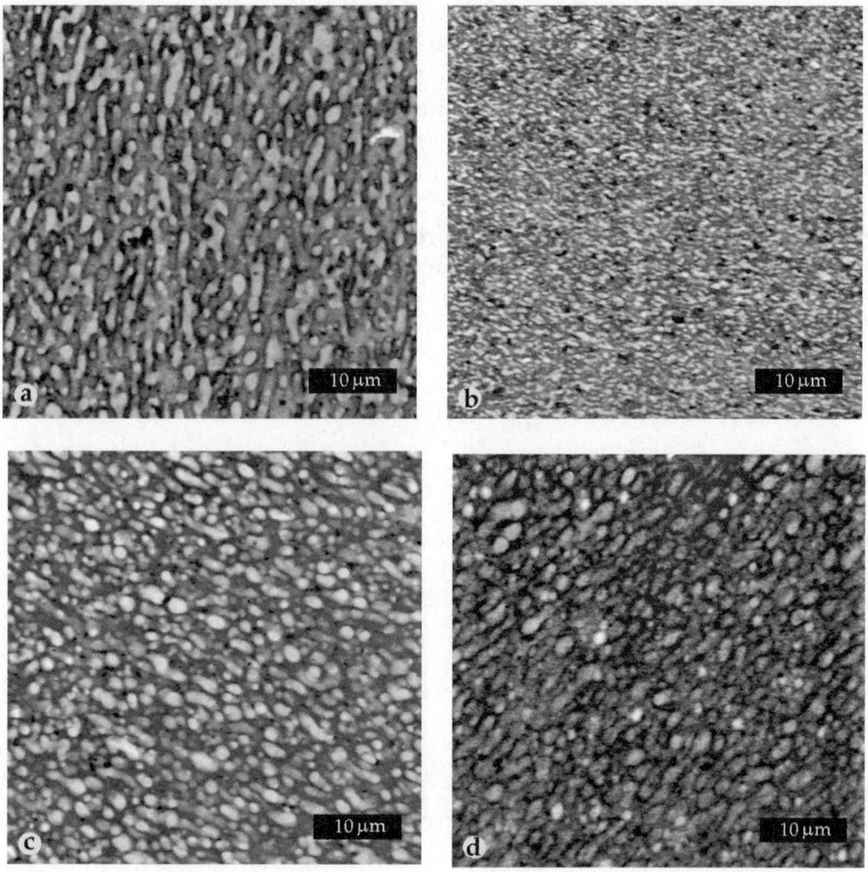


**Figure 14.5** STEM micrographs of 60:40 NR:EPDM gum blend vulcanizates (NR stained with osmium tetroxide and appears dark) showing effect of modifier.

to follow similar trends, although identification of the separate phases by STEM, or even TEM, has proved to be extremely difficult.

## 14.6 INTERACTION OF MODIFIED EPDM AND CARBON BLACK

Normally, only relatively low levels of interaction are developed between EPDM and filler particles such as carbon black. This behaviour contrasts strongly with NR which is able to form a high degree of association with carbon black [21]. This reflects the paucity of reactive sites on the EPDM polymer chain. EPDM modification by each of the sulphur donors has been shown (Figure 14.6) to have a significant effect upon the level of interaction between EPDM and carbon black, as determined by bound rubber and volume swelling measurements. It has been found that for broadly similar levels of modification, the amount of polymer bound to the carbon black after swelling in toluene is more than doubled after modification by BAPD in comparison with an unmodified control masterbatch compound. While the effect of DTDC modification is somewhat less pronounced, modification with DTDM results in approximately 75% of the polymer becoming bound to the carbon black. Volume swelling measurements, which indicate how tightly the polymer is bound to carbon black, also show an increase in the volume fraction of rubber in the swollen gel, $V_r$, for all of the modified EPDMs, with the greatest increase being given by the DTDM modification.

## 14.7 PHYSICAL PROPERTIES

The combined effects of the different sulphur donor modifying agents can be seen in the physical properties of black-filled (50 phr N660) 60:40

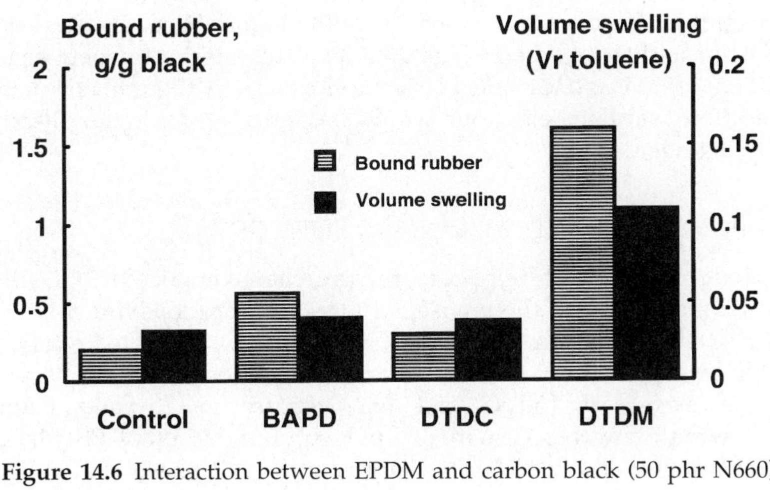


**Figure 14.6** Interaction between EPDM and carbon black (50 phr N660).

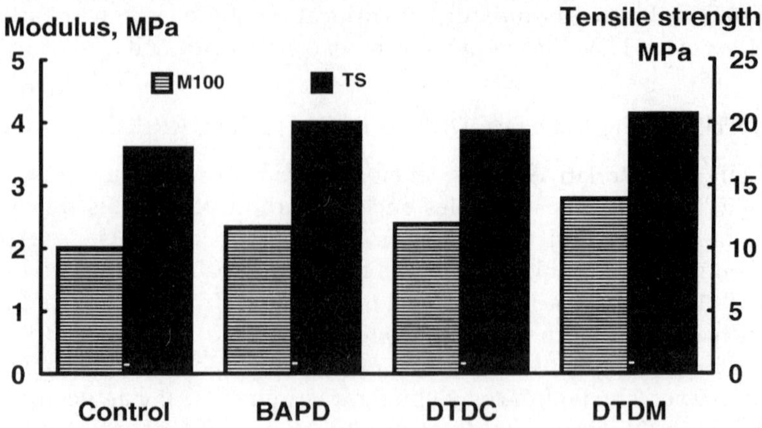


**Figure 14.7** Tensile properties of black-filled (50 phr N660) NR/EPDM blends (2.0 phr S/0.6 phr CBS, $t_{max}$/150 °C).

NR:EPDM blends cured with 2.0 phr sulphur/0.6 phr CBS to optimum cure level ($t_{max}$) at 150 °C. As seen in Figure 14.7, all of the modifiers produce an increase in modulus and tensile strength when compared with the unmodified EPDM blend. The high modulus of the DTDM-modified EPDM blend is likely to be a result of the very high level of interaction developed between the modified EPDM and carbon black. The effect of the BAPD modification on the more 'dynamic' of physical properties such as Goodrich heat build-up (ISO 4666/3, 1982) can be seen in Figure 14.8. At three different loadings of N660 carbon black in the EPDM masterbatch, (NR masterbatch loading constant at 50 phr N660), the effect of the modification is to reduce heat build-up significantly. As expected this effect is most noticeable at the higher black loadings in the EPDM masterbatch, where a greater level of interaction between the modified EPDM and carbon black would exist. Although not demonstrated here, similar behaviour would be anticipated for the other two modifying agents.

## 14.8 HIGH TEMPERATURE CURING EFFECTS

Knowledge of the high temperature cure characteristics of NR/EPDM blends is important as the envisaged target applications are commonly produced at temperatures in excess of 170 °C. It is expected that a reasonable loss of crosslinking or reversion should occur at cure temperatures in excess of 150 °C and on overcure in a blend using a 'conventional' curative system (2.0 phr sulphur/0.6 phr CBS). Indeed, this is what is seen by $^1$H NMR analysis of the NR phase of NR/EPDM

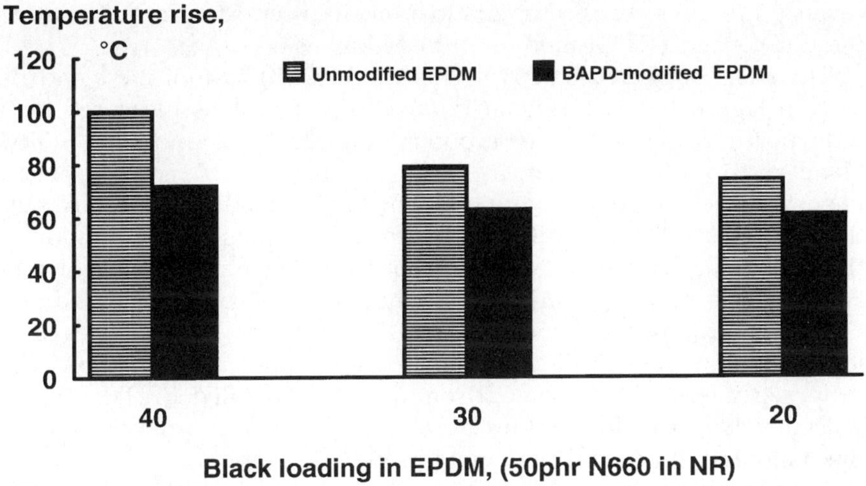


**Figure 14.8** Effect of black loading on heat build-up in NR/EPDM blends (2.0 phr S/0.6 phr CBS, $t_{max}$/150 °C).

gum blends (Figure 14.9). For the modified EPDM blends, several important features can be seen. For all of the modified EPDM blends, an increase in crosslink density in the NR phase is seen. This is thought to be a result of utilization by the NR of unreacted sulphur donor or active sulphur donor fragments from the modified EPDM phase. However, severe reversion on overcure at 175 °C in the NR phase is observed for both the unmodified and BAPD-modified EPDM blends, although

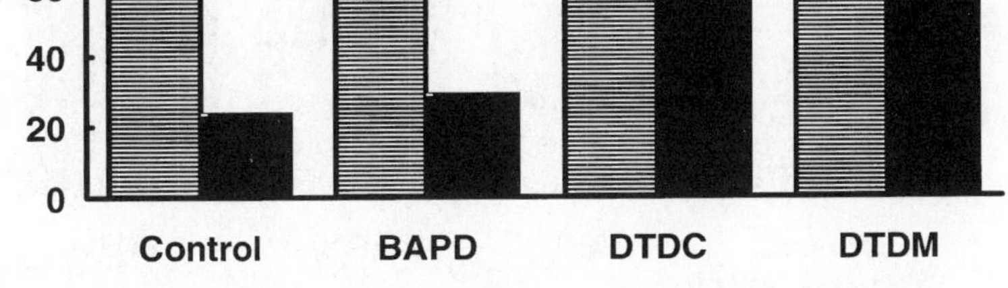


**Figure 14.9** Effect of cure temperature on NR crosslink density (2.0 phr S/0.6 phr CBS).

reversion in the NR phase occurs to a much lesser extent in the cases of the DTDC- and DTDM-modified EPDM blends.

The effect of overcure at 175 °C in the EPDM phase of the blends is seen in Figure 14.10. Reversion in the EPDM phase, as measured by a reduction in crosslink density, is observed for the BAPD-modified EPDM blend and to a lesser extent in the unmodified EPDM 'control' blend. However, such loss of crosslinking in the EPDM phase is absent in either the DTDC- or DTDM-modified EPDM blends. A possible explanation of this behaviour may be made if the proposed reaction pathway, shown in Figure 14.1, is examined. In the case of both DTDC- and DTDM-modified EPDM the proposed polymer pendant group intermediate is believed to lead predominantly to di-sulphidic crosslinks upon vulcanization. The proposed intermediate derived from a BAPD-modified EPDM can, on vulcanization, lead to sulphur crosslinks of a higher order which are thermally less stable and thus more prone to reversion.

Mixed modification systems have demonstrated that 'modifier specific' attributes can be combined. Thus a mixture of BAPD and DTDC gives a higher level of crosslinking in the EPDM phase of a blend with NR cured at high temperature than does BAPD alone. In addition to this, an improvement in blend phase morphology intermediate to that of either BAPD or DTDC alone is observed. Increasing the overall curative level in the blend will also help to offset the effects of reversion when using a conventional cure system for NR/EPDM blends cured at high cure temperature, and where overcuring is a likely occurrence.

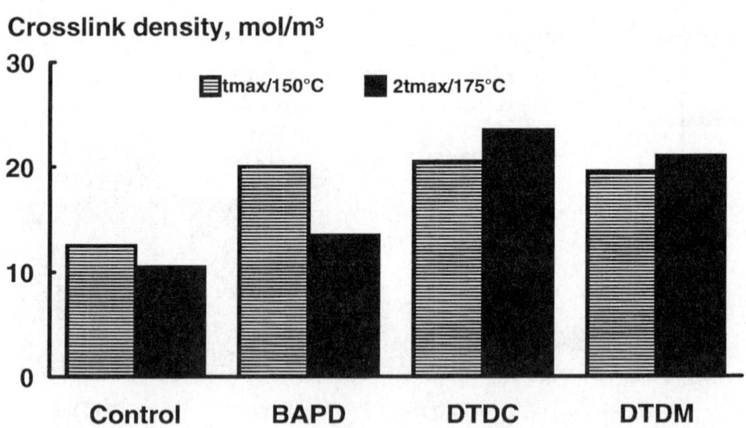


**Figure 14.10** Effect of cure temperature on EPDM crosslink density (2.0 phr S/ 0.6 phr CBS).

## 14.9 MODIFICATION OF DIFFERENT EPDM GRADES

To demonstrate the generality of the modification procedure, several different grades of EPDM have been successfully modified with sulphur donors. EPDM grades with 'very high' levels of unsaturation (>9 wt%) are to be favoured although, as discussed above, this level of unsaturation corresponds to only some 2 mole% and so is in fact still very low in comparison to NR. These grades commonly have medium (50–60 wt%) ethylene contents. EPDM grades which use the commonly available termonomers, ethylidene norbornene (ENB), dicyclopentadiene (DCPD) and 1,4-hexadiene have all been successfully modified, although the best overall level of properties are obtained from ENB grades. Oil extended (OE) EPDM grades have also been successfully modified, although the mixing conditions stipulated above may be more difficult to achieve because of the very high oil contents of some OE EPDMs.

Modification of the very high molecular weight OE EPDM grades allows a higher level of physical crosslinking in the EPDM phase of a blend to be achieved which is beneficial to properties such as compression set. Increased levels of interaction between high molecular weight grades of EPDM and carbon black can also be developed during 'reactive mixing', again as a consequence of their longer chain lengths.

## 14.10 CONCLUSION

A new technology has been developed whereby NR is cross-blended with a modified EPDM to give a blend resistant to ozone attack without requiring conventional antiozonant protection. This new technology has been shown to be suitable in 60:40 NR:EPDM blends for a number of different applications. Modification of EPDM in the internal mixer has been achieved with three commercially available sulphur donors as part of a full masterbatch mixing stage. A range of different EPDM grades has been successfully modified, although the best results are obtained from EPDM with a high level of ENB. The modification has been shown by swollen-state NMR methods to give an improved level of crosslinking in the EPDM component of a blend with NR under a wide range of cure conditions. The modification also results in an enhanced level of interaction between EPDM and carbon black, although the modification itself has been shown to be more efficient in white-filled compounds (Chapter 16). For one of the sulphur donors, a reduction in phase size of the blend is also produced. At high cure temperatures, 170–180 °C, the preferred EPDM modification system has been shown to be a mixture of two of the sulphur donors. The effects of the EPDM modification procedure in a blend with NR are beneficial in terms of producing improvements in a broad range of physical properties.

## 14.11 REFERENCES

1 Morrissey, R.T. (1971) Halogenation of ethylene propylene diene rubbers. *Rubber Chemistry and Technology*, **44**(4), 1025–1042.
2 Baranwal, K.C. and Son, P.N. (1974) Co-curing of EPDM and diene rubbers by grafting accelerators onto EPDM. *Rubber Chemistry and Technology*, **47**(1), 88–99.
3 Hashimoto, K., Miura, M., Takagi, S. and Okamoto, H. (1976) Co-vulcanization of EPDM and NR 1. *International Polymer Science and Technology*, **3**, T84–T87.
4 Hopper, R.J. (1976) Improved co-cure of EPDM-polydiene blends by conversion of EPDM into macromolecular cure retarder. *Rubber Chemistry and Technology* **49**(1), 341–352.
5 Mastromatteo, R.P., Mitchell, J.M. and Brett, Jr., T.J. (1971) New accelerators for blends of EPDM. *Rubber Chemistry and Technology*, **44**(4), 1065–1079.
6 Woods, M.E. and Mass, T.R. (1975) Fundamental considerations for the covulcanization of elastomer blends, in *Copolymers, Polyblends and Composites – Advances in Chemistry Series* (ed. N.A.J. Platzer), vol. 142, pp. 386–398.
7 Coran, A.Y. (1987) *Making rubber blends of diene rubber and EPR or EPDM*. US Patent 4,687,810, 18 August, Monsanto Company, St Louis.
8 Shell Chemical Company (1965) *Technical Bulletin SCR*, 65–160.
9 Andrews, E.H. (1967) Resistance to ozone cracking in elastomer blends. *Rubber Chemistry and Technology*, **40**(2), 635–649.
10 Sutton, M.S. (1964) Blends of Royalene with other polymers. *Rubber World*, **149**(5), 62–68.
11 Corish, P.R. (1967) Fundamentals of rubber blends. *Rubber Chemistry and Technology*, **40**(2), 324–340.
12 Rhener, Jr., J. and Wei, P.E. (1969) Heterogeneity and crosslinking of elastomer blends. Investigations of chlorobutyl with highly unsaturated elastomers. *Rubber Chemistry and Technology*, **42**(2), 985–999.
13 Kerrutt, G., Blumel, H. and Weber, H. (1969) Untersuchungen zur covulkanisation. *Kautschuk Gummi Kunststoffe*, **22**(8), 413–418.
14 Woods, M.E. and Davidson, J.A. (1976) Fundamental considerations for the covulcanization of elastomer blends. II. Lead oxide-activated cures of NBR-EPDM blends. *Rubber Chemistry and Technology*, **49**(1), 112–117.
15 Shershnev, V.A. (1982) Vulcanization of polybutadiene and other hydrocarbon elastomers. *Rubber Chemistry and Technology*, **55**(3), 537–574.
16 Gardiner, J.B. (1968) Curative diffusion between dissimilar elastomers and its influence on adhesion. *Rubber Chemistry and Technology*, **41**(1), 1312–1328.
17 Brown, P.S. and Tinker, A.J. (1996) The effect of modifying EPDM in the crosslink distribution of NR/EPDM blends. *Journal of Natural Rubber Research*, submitted for publication.
18 van Duin, M., Krans, J.C.J. and Smedinga, J. (1993) Covulcanization and ozone resistance of NR/EPDM blends. *Kautschuk Gummi Kunststoffe*, **46**(6), 445–451.
19 von Hellens, C.W. (1989) Some considerations for the application of EPDM in tyre components. Paper No. 40, ACS Rubber Division Meeting, Detroit, Michigan, USA, October 17–20.

20 Zamboni, V., Flisi, U. and Giunchi, G. (1971) Crosslink density evaluation of EPDM vulcanizates. *Rubber Chemistry and Technology*, **44**(4), 1109–1129.

21 Meissner, B. (1974) Theory of bound rubber. *Journal of Applied Polymer Science*, **18**, 2483–2491.

# 15

# Compounding NR/EPDM blends for tyre sidewalls

*Stuart Cook*
*Tun Abdul Razak Research Centre, MRPRA, UK*

## 15.1 INTRODUCTION

A tyre sidewall provides the physical connection between a vehicle's wheel and the tyre tread: it transmits power and braking force to the tyre tread which is itself in contact with the road surface. The sidewall also provides a significant part of a vehicle's suspension and plays an important role in the general handling of the vehicle on the road. As can be seen in Figure 15.1, the sidewall is a remarkably thin and seemingly fragile part of the tyre considering its function.

The trend today is for passenger vehicles to be fitted with radial ply tyres with the requirement that they continue to meet ever-increasing performance criteria necessary to match the rising capabilities of modern cars. This has led to tyres which are lighter in weight, have lower profiles and a wider tread to give improved grip and better high speed handling. These requirements have put an even greater demand on the sidewall compound as the flex area is concentrated into an even smaller region of the tyre. In addition, the expected lifetime of a modern tyre is greater than ever before at around 80 000 kilometres. The essential property requirements [1] of a modern sidewall compound are thus to have a high resistance to flex cracking with good weathering and fatigue resistance. Other desirable features that the sidewall compound should exhibit are good processability and good adhesion to carcass compounds during manufacture and adequate abrasion resistance during service.

*Blends of Natural Rubber*. Edited by Andrew J. Tinker and Kevin P. Jones.
Published in 1998 by Chapman & Hall, London. ISBN 0 412 819406

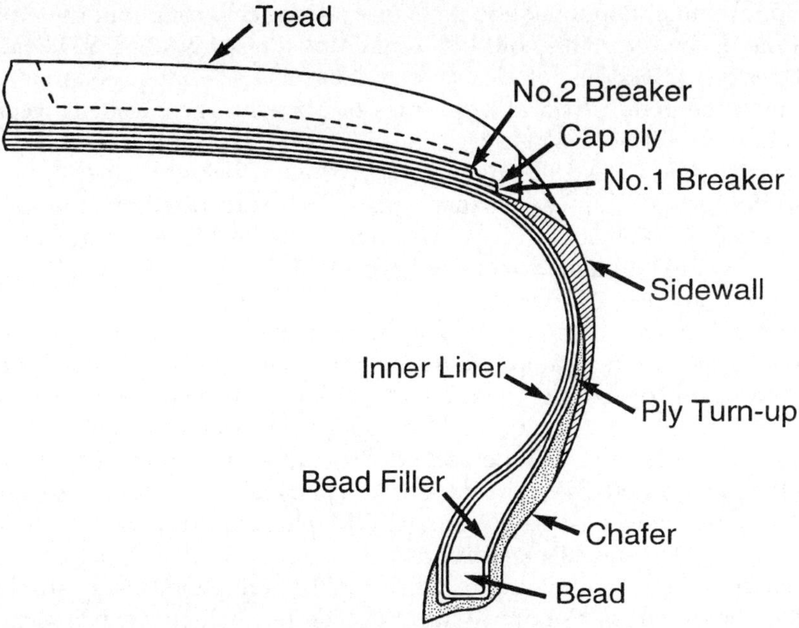


**Figure 15.1** Structure of a tyre.

Current tyre black sidewall compounds are generally based on blends of natural rubber (NR) with polybutadiene (BR) and contain antioxidants, antiozonants and waxes to protect against cracking and environmental weathering. During the lifetime of the tyre the antiozonants and waxes present are brought to the surface which discolours the sidewalls and, as the protectants become depleted, the sidewalls harden and crack. The appearance of the tyre sidewall deteriorates quite noticeably long before cracking occurs and continues to do so throughout the lifetime of the tyre. For a car, the appearance of the tyre contributes to the overall appearance of the vehicle and so is of some importance to the manufacturer. A tyre sidewall built using a blend of NR with ethylene-propylene-diene monomer (EPDM) would be protected against attack by ozone (Chapter 14) and, having no need for antiozonant, would maintain its initial 'new' appearance and would thus be advantageous.

Truck tyres usually undergo more than one retread in their lifetime and so it is important that the sidewall should be able to perform adequately for the entire life of the tyre. Depletion of antidegradants in the sidewall can become a serious problem even after only the first or second retread. Thus, if the ultimate mileage target for a truck tyre of one million miles [2] is ever to be realized, the tyre would be expected to undergo three or four retreads and would be dependent upon the sidewall compound maintaining adequate ozone resistance throughout the entire lifetime of the tyre. This could be offered by a 60:40 NR:EPDM blend.

Over the past three decades several attempts [3–5] have been made to combine the good physical properties of NR with the excellent weatherability of EPDM for use in tyre sidewalls, all with scant success. Originally, NR/EPDM blends were perceived as being of use primarily in white sidewalls [6] where ozone resistance was required and the use of antiozonants that stain or bloom was unacceptable. However, as interest in the NR/EPDM blend system developed, black sidewalls have become the major target.

Whilst it has been shown [7] that the proportion of EPDM required to give adequate protection to NR against attack by ozone is about 40 wt%, the physical properties of such blends, although strongly influenced by the NR content, have not proved acceptable for black tyre sidewalls. As discussed in Chapter 14, the reasons for this can be attributed to the fundamental problems of low crosslink density in the EPDM phase, poor crosslinking between the NR and EPDM phases and poor interaction between EPDM and black. Solutions to these problems are based on a 'reactive mixing' technology in which EPDM is modified using sulphur donors before cross-blending with NR. The technology, which was introduced in Chapter 14, is developed in this chapter in the context of tyre sidewalls.

## 15.2 PROPERTIES OF COMPARABLE NR/BR AND NR/EPDM BLEND FORMULATIONS

A typical tyre sidewall compound is a 50:50 NR:BR blend containing carbon black (50 phr N660) and oil (10 phr) and a variety of other additives, including antiozonants. Such a compound would be expected to have a tensile strength of over 20 MPa and a modulus at 100% strain (M100) value of about 1.5 MPa (Figure 15.2) when cured at 150 °C to optimum cure ($t_{max}$). In comparison, a 60:40 NR:EPDM (Vistalon 6505) blend having the same black and oil loading but no antiozonants, cured under similar conditions, has a slightly reduced tensile strength but a significantly increased modulus (M100). Use of EPDM modified with 1 phr of the sulphur donor *bis*-alkylphenoldisulphide (BAPD) in a blend with NR, although improving the tensile strength of the blend, produces a further increase in compound modulus.

Using cure conditions more usual for car tyre production, a cure temperature of 170 °C for a duration of twice optimum cure ($2 \times t_{max}$), the tensile strength of the modified-EPDM blend (Figure 15.3) compares more favourably with that of the NR/BR control compound but a greater differential in moduli values is apparent. The higher modulus of the NR/modified-EPDM blend is believed to result in part from the high cross-

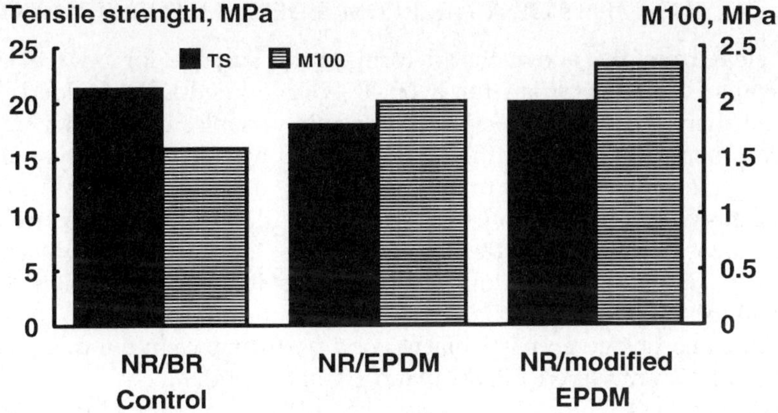


**Figure 15.2** Tensile properties of black-filled blends (50 phr N660, 10 phr oil, $t_{max}$/150 °C).

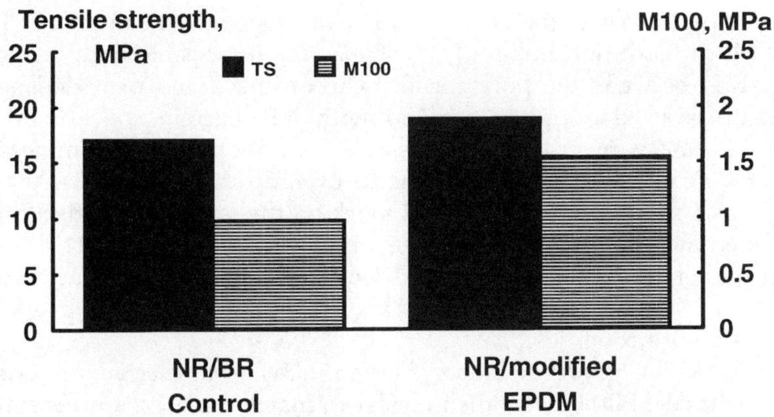


**Figure 15.3** Tensile properties of black-filled blends (50 phr N660, 10 phr oil, $2 \times t_{max}$/170 °C).

link density of the dominant NR phase in the NR/EPDM blend, the higher $T_g$ of EPDM compared with BR and from the higher level of interaction developed between modified-EPDM and carbon black. A significantly higher modulus in a sidewall compound is undesirable as it might be expected to produce a greater stress concentration in the shoulder region of the tyre, thus increasing the risk of failure at this point.

## 15.3 DEVELOPMENT OF A NR/EPDM SIDEWALL FORMULATION

It is clear from the above that a formulation suitable for NR/BR tyre sidewalls is inappropriate for a 60:40 NR:modified-EPDM blend. Although there are a large number of possible variables to consider in the development of a NR/modified-EPDM blend appropriate for tyre sidewalls, some simplifications may be made.

The polymer blend ratio can be set at 60:40 NR:EPDM to ensure adequate ozone resistance of the compound [7]. A particular EPDM grade (Polysar 585) had been adopted initially as it has a very high level (~11%) of ethylidene norbornenyl (ENB) unsaturation, it is readily available and has given particularly good performance in the more basic studies of NR/modified EPDM blends (Chapter 14).

Investigations into the choice of EPDM modification system (Chapter 14) suggested that a combination of *bis*-alkylphenoldisulphide (BAPD) and dithiodicaprolactam (DTDC) were to be preferred. Whilst BAPD offers an improved phase morphology in a blend, its effect in improving EPDM crosslink density at high cure temperatures is limited. Modification of EPDM with DTDC, however, shows a greater resistance to reversion in a blend than does BAPD. Although modification of EPDM with 4,4′-dithiodimorpholine (DTDM) has made possible high levels of interaction between the polymer and carbon black and provided good reversion resistance and in a blend with NR, current concern in the rubber industry regarding the possible formation of nitrosamines [8] during vulcanization would appear to exclude the commercial use of DTDM as a modifying agent. Initial work on the level of modifier to use suggested overall levels up to about 2 phr, with there being no benefits to using higher levels of modifier. Subsequent work showed that a combination of 0.8 phr BAPD with 0.4 phr DTDC was the most appropriate for a range of properties.

A reinforcing grade of carbon black (N339) was selected for use in NR/modified-EPDM sidewalls in order to match the abrasion resistance of current NR/BR sidewalls containing N660 carbon black. The EPDM modification procedure requires a separate masterbatch then cross-blend mixing stages which permit different black loadings to be used in the two phases of the blend. This possibility may bring additional benefits not available from a single-stage mixing procedure, although identification of appropriate black loadings for each masterbatch would be a complex matter involving considerable experimentation. The choice of a paraffinic oil rather than a naphthenic or aromatic oil has been suggested elsewhere [9], but identification of the most appropriate loading would again involve considerable experimentation.

A suitable cure system based on sulphur and CBS has been identified after comparisons with a typical NR/BR sidewall compound and by using crosslink density measurements obtained by the swollen-state NMR spectroscopy [10] of gum blend vulcanizates.

Of the variables identified above, perhaps the three most difficult to assess without further experimentation are black loading in each of the separate masterbatches and the overall oil level in the blend. The most efficient way of assessing three such variables is to use an experimental design approach [11].

## 15.4 EXPERIMENTAL DESIGN APPROACH

The scope of a central rotatable composite experimental design, incorporating the three variables described above, is shown in Table 15.1. Initial studies had indicated that the centre point chosen for the design (25 phr N339 in the EPDM, 50 phr N339 in the NR and 20 phr oil in the blend overall), having the design coordinate (0, 0, 0), gave reasonable properties and would allow a comprehensive range of black and oil loadings to be included in the design. Such a three-variable design requires 20 experimental compounds (including six centre-point replicates) to be prepared, each requiring two masterbatch compounds, a cross-blending stage then a finalizing stage in which curatives are added. Compound preparation was performed by using a 00C Banbury mixer (3 litre batch size) with finalizing of a portion of the cross-blend being performed on a two-roll mill. Each design compound was vulcanized at 175 °C for 12 minutes (15 minutes for heat build-up test pieces) which is typical of the cure conditions used in factory passenger tyre production. The design compound formulations are shown in Table 15.2.

The mixing procedure followed for the design compounds is clearly very important; it is essential to mix as consistently as possible to avoid the inadvertent introduction of any uncontrolled variables to the design. Since the EPDM modification procedure has been shown to be dependent on mixing conditions, a particular mixing procedure was chosen in which the black:oil ratio was maintained at 10:1 for each masterbatch. Any additional oil required by the blend was added on cross-blending.

**Table 15.1** Scope of experimental design

| | *Design coordinate* | | | | | *Variable* |
|---|---|---|---|---|---|---|
| | *−1.68* | *−1.0* | *0* | *+1.0* | *+1.68* | |
| $x$ | 0 | 10.12 | 25 | 39.88 | 50 | black loading in EPDM, |
| $y$ | 25 | 35.12 | 50 | 64.88 | 75 | black loading in NR, phr |
| $z$ | 5 | 11.07 | 20 | 28.93 | 35 | total oil in blend, phr |

**Table 15.2** Design compound formulations

| *EPDM m/b* | *phr* | *NR m/b* | *phr* |
|---|---|---|---|
| Polysar 585 | 100 | SMR L | 100 |
| BAPD | 0.8 | N339 | y |
| DTDC | 0.4 | Paraffinic oil | y/10 |
| N339 | x | | |
| Paraffinic oil | x/10 | | |

zinc oxide (3 phr) and stearic acid (2 phr) common to both

| | |
|---|---|
| Cross-blend | To achieve a 60:40 NR:EPDM polymer ratio and add remaining oil $z - (x + y)/10$ |
| Cure system | 2.0 phr S/1.0 phr CBS, cured for 12 min/175 °C |
| Scale | 00C Banbury (3 kg batch size) |

The temperature profile experienced by each EPDM masterbatch was kept as consistent as possible by typically mixing to a batch temperature of 150 °C and then for a further 1.5 minutes. Subsequent GPC/UV analysis of the EPDM masterbatches indeed showed very similar levels of modification (~0.25 wt% bound modifier) to have been attained in all of the mixes. Although the mixing procedure described ensured a high degree of consistency to be achieved between compounds it was recognized as being 'unrealistic' to add quantities of oil at the cross-blending stage as this would be unlikely to be followed in practice.

The test results from the design compounds must be seen and analysed as a whole rather than individually. Regression analysis of raw physical test results for each compound enables a response equation to be constructed for each property of interest having the following form:

$$\begin{aligned}\text{Property value} = {} & C_0 + C_x V_x + C_y V_y + C_z V_z + C_{xx} V_x V_x + C_{yy} V_y V_y \\ & + C_{zz} V_z V_z + C_{xy} V_x V_y + C_{xz} V_x V_z + C_{yz} V_y V_z \\ & + C_{xyz} V_x V_y V_z\end{aligned}$$

where $V$ corresponds to the design coordinate for each variable, $x$, $y$ or $z$. The response equations are modified by using the t-test [12] to exclude non-valid terms.

A wide range of physical properties were measured for all of the design compounds and response equations were derived for each property. However, as discussed above, not all properties are of equal importance for a tyre compound. Of those measured, four key properties were identified; modulus (M100), tear strength (trouser tear), abrasion resistance (Akron abrasion volume loss) and Goodrich heat build-up.

**Table 15.3** Mean deviations from predicted properties

| *Experimental property value* | *M100 (MPa)* | *Trouser tear strength (N/mm)* | *Akron abrasion, volume loss (mm³)* | *HBU (°C)* |
|---|---|---|---|---|
| Design experiment mixing procedure | −5% | −28% | −9% | −3% |
| Normal mixing procedure | +6% | −20% | −22% | −3% |

These are also the type of properties in which NR/EPDM blends had previously been shown to be deficient.

An important part of a design is its ability to predict, with a high degree of certainty, the properties of compounds that lie within the design limits. Four compounds were used to check the validity of the design in this respect. Two of these compounds were mixed in the same way as the design compounds and two were mixed in a more 'normal' manner, i.e. adding most of the oil to the NR masterbatch to avoid any oil addition at the cross-blending stage. This would enable the effect of the mixing procedure on predicted compound properties to be assessed. A summary of the deviations from predicted results of the test compounds for the four key properties is shown in Table 15.3.

Good agreement is shown between predicted and experimental M100 values of compounds mixed in the same manner as the design compounds, but somewhat better than predicted abrasion and heat build-up performances are achieved. This improvement becomes more apparent for compounds mixed in the more 'normal' manner. Perhaps unsurprisingly, it would appear that the stress–strain behaviour of the compounds depends upon the manner in which the compound is mixed, as M100 is increased somewhat for the compounds that were mixed in the more 'normal' manner. Tear strength, as measured by trouser tear, is however worse than predicted by the design, although improvements are seen following the 'normal' mixing procedure.

To identify an optimum or 'preferred' formulation, the design response equations for the key properties above are combined to produce an overall composite desirability function [11]. Limits of 'desirability' in the form of property constraints are imposed on the four key properties (Table 15.4). A 'preferred' compound is identified as one having the best overall properties, and thus the highest 'desirability' in the above key categories, whilst meeting all of the imposed property constraints. Since it has been demonstrated that a more 'normal' mixing procedure for the blend is likely to result in an improvement over the properties predicted by the design, some additional latitude was given to the limits imposed on tear strength, abrasion resistance and heat build-up. The preferred

**Table 15.4** Key property restraints

| *Property* | *Lower limit* | *Optimum* | *Upper limit* |
|---|---|---|---|
| M100, MPa | 0.9 | 1.1 | 1.2 |
| Trouser tear, N/mm | 5.5 | – | >10 |
| Akron abrasion, volume loss, $mm^3$ | <80 | – | <100 |
| Heat build-up, °C | <80 | – | <116 |

**Table 15.5** Optimized formulation identified from preferred key property values

| *Design coordinate* | *EPDM black* | *NR black* | *Oil* |
|---|---|---|---|
| Design units | 0.4 | −0.2 | 0.4 |
| phr | 31 | 47 | 23.6 |

formulation identified by the composite desirability function is shown in Table 15.5.

Blends having the preferred formulation were prepared in mixers fitted with tangential (Banbury) type and intermeshing (Intermix) type rotors up to relatively large laboratory scale (~33 kg), thus demonstrating that the reactive mixing EPDM modification technology can be successfully applied to larger-scale mixing. Although satisfactory levels of EPDM modification were achieved in all cases, it was apparent that 'over-modification' could occur if the modified-EPDM masterbatch is allowed to remain at temperatures in excess of 150 °C for more than about eight minutes. The manifestation of such 'over-modification' is the development of a very high level of interaction between the EPDM and carbon black which renders the compound crumbly and very difficult to process. Such problems are likely only to be encountered in large-scale mixing in a factory where delays in production processes downstream of the dump mill can sometimes occur and lead to freshly mixed compound having to be held back for a period of time. In a factory operation the time taken to process and cool a large batch on a mill is usually dictated by the mix cycle time and is usually much less than eight minutes, although for a dump extruder processing several consecutive batches, the compound may remain at a relatively high temperature for longer than on a mill.

The general physical properties of NR/modified-EPDM blends having the preferred formulation are satisfactory, with several properties being considerably better than predicted by the design, and are comparable with typical NR/BR sidewall compounds. However, two important properties, tear strength and fatigue life, are well below the levels achieved in typical NR/BR sidewall compounds.

It has previously been shown [13] that incorporation of a high molecular weight EPDM grade in a blend with NR is of considerable benefit in improving fatigue life and crack growth resistance and also for improving adhesion of the blend to other components of a tyre based on highly unsaturated elastomers. Polysar 5875 is an EPDM grade having a particularly high molecular weight and a high level of oil-extension (100 phr) together with a relatively high level of ENB unsaturation ($\sim$9 wt%). Whilst modification by sulphur donors of this grade is problematic because of its high oil loading and subsequently reduced mix viscosity, part replacement of the non oil-extended EPDM (Polysar 585) with oil-extended EPDM, i.e. a 70:60 Polysar 585:Polysar 5875 blend, has been shown to be beneficial in a number of respects. Incorporation of oil into the blend is made considerably more efficient whilst maintaining acceptable levels of modification. Higher EPDM physical crosslink densities were also attainable for a given level of curative because of the very high molecular weight of Polysar 5875. However, such a high molecular weight of Polysar 5875 can also lead to higher levels of interaction with carbon black being developed which either contribute to a higher modulus in a blend with NR or, if 'over-modification' occurs, result in processing difficulties. Although fatigue life and tear strength of a preferred formulation blend is improved by the incorporation of Polysar 5875, these properties remain inferior to those found in the NR/BR sidewall control compounds.

## 15.5 CRACK GROWTH TESTING

Measurements of tear strength and De Mattia crack growth rates may give an indication of actual 'in service' performance in some circumstances, but it has been shown that results from this test can often be misleading or even incorrect when studying sidewall compounds [14]. Collaboration with a major EPDM manufacturer allowed access to crack growth testing facilities that have been shown to correlate well with actual tyre testing [14]. Subsequent discussions with several tyre manufacturers indicated crack growth resistance to be the prime consideration in assessing the performance of a tyre sidewall compound. Although good fatigue life is desirable, a low modulus is not necessarily desirable and a slightly higher modulus may be preferable as it imparts a better 'ride' and handling to a vehicle.

Crack growth testing of the design compounds using the facilities of the EPDM manufacturer showed that if crack growth resistance alone was to be considered, then a compound having a low level of black in the

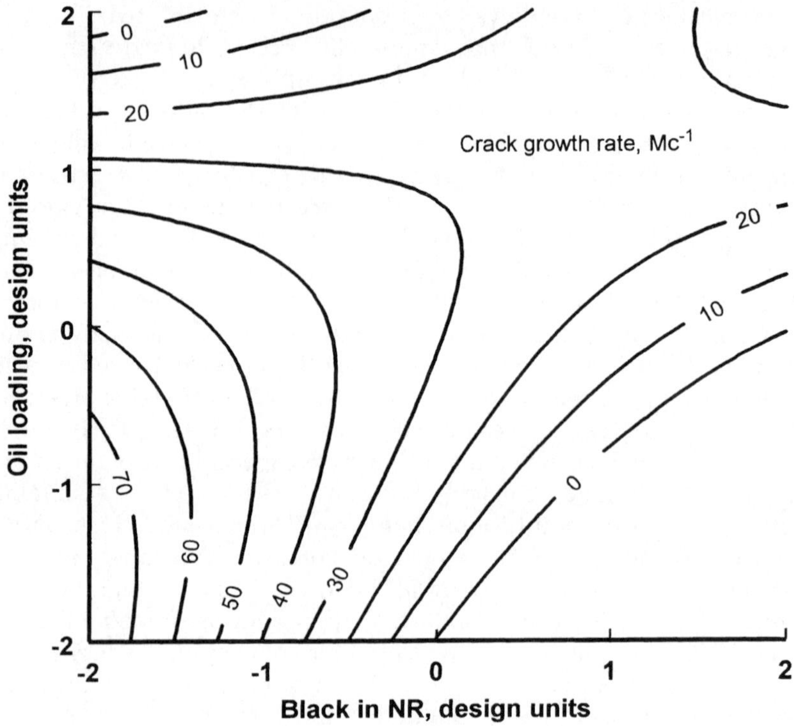


**Figure 15.4** Effect of changing NR masterbatch black and oil loadings on vulcanizate crack growth rate.

EPDM masterbatch, a high level of black in the NR masterbatch and lower levels of oil overall should be favoured. The effects of changing the black loadings in each of the masterbatches can be observed in Figures 15.4 and 15.5. The relative crack growth rates appear to increase markedly if the EPDM black loading is increased and also if the NR black loading is reduced. Although not so dramatic, increased levels of oil are also often associated with increased crack growth rates. The effect of following a more 'normal' mixing procedure than that adopted for the design resulted in further improvements to crack growth resistance, as had been demonstrated previously for a number of other properties.

Additional factors were found to improve crack growth performance in compounds having the type of black distribution described above. A small reduction in the overall level of curatives is beneficial [15], so too is the addition of antioxidant [16]. Although it was originally believed that EPDM would help to protect NR against oxidation as well as against ozone degradation, this has not proved to be sufficient. Addition of antioxidant must be made in the final stages of compounding after EPDM

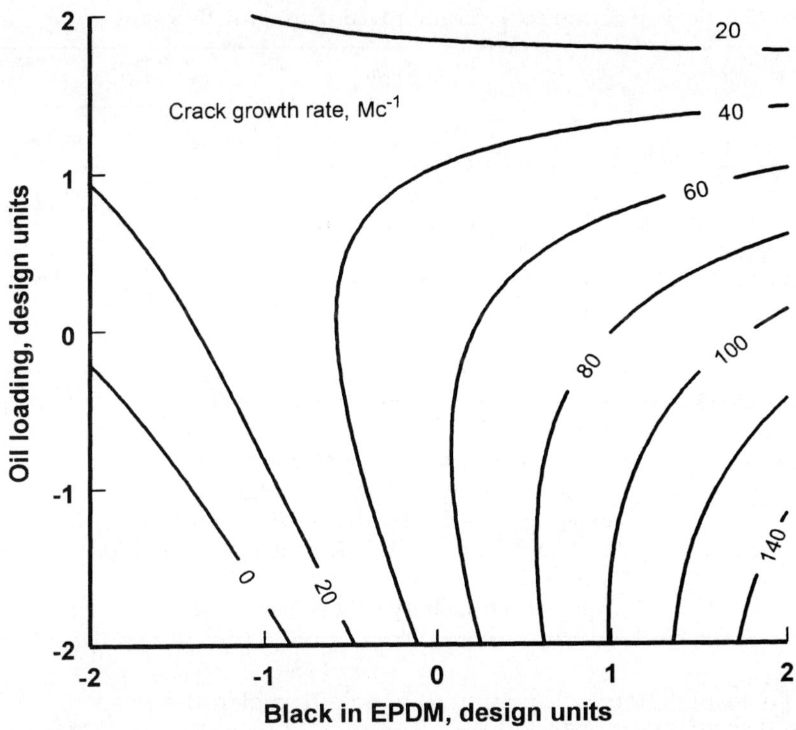


**Figure 15.5** Effect of changing EPDM masterbatch black and oil loadings on vulcanizate crack growth rate.

modification to avoid interference with the morphological improvements sought at the cross-blending stage. An additional benefit of adding antioxidant to the blend is seen in a greatly improved abrasion resistance. This is not altogether surprising as abrasion is a mechanico-oxidative process, similar in some respects to crack propagation. Rather surprisingly, no benefits were found on changing from N339 to less reinforcing grades of black (to give a lower modulus), such as comparable levels of N660 on its own, or a N375/N660 mixture.

The revised formulation (compound A) for improved crack growth resistance is shown in Table 15.6. The same EPDM modification conditions as described earlier are recommended, that is, the EPDM masterbatch is mixed to 150 °C and the batch temperature maintained at between 150 °C and 165 °C for between 1 and 1.5 minutes. Satisfactory mixing of this formulation has been achieved in a Farrel F50 Banbury fitted with Synchronous Technology (ST) rotors (33 kg batch size). Physical properties of a blend having this formulation are shown in Table 15.7. It can be seen that in many respects the revised NR/modified-

**Table 15.6** Revised formulation for improved crack growth resistance

| *EPDM m/b* | *phr* | *NR m/b* | *phr* | |
|---|---|---|---|---|
| | | | *m/b A* | *m/b B* |
| Polysar 585 | 70 | SMR L ($M_L$ 1 + 4 = 60) | 100 | 100 |
| Polysar 5875 | 60 | N339 | 60 | 50 |
| BAPD | 0.8 | Paraffinic oil | 7 | 7 |
| DTDC | 0.4 | | | |
| N339 | 10 | | | |

Zinc oxide (3 phr) and stearic acid (2 phr) common to both

| | |
|---|---|
| Cross-blend | To achieve a 60:40 NR:EPDM polymer ratio<br>Compound A – EPDM m/b + NR m/b A<br>Compound B – EPDM m/b + NR m/b B |
| Cure system | 1.8 phr S/0.75 phr CBS + 1.0 phr TMQ, cured for 12 min/175 °C |
| Scale | Up to No. 11 Banbury (200 kg batch size) |

EPDM formulation (compound A) is able to match the properties of a typical 50:50 NR:BR sidewall compound mixed using 50 phr of N660. For comparison, physical properties of a 50:50 NR:BR compound using

**Table 15.7** Predicted and experimental physical properties of the preferred NR/EPDM blend and NR/BR control compounds

| *Property* | *NR/EPDM* | | *NR/BR* | |
|---|---|---|---|---|
| | *Predicted* | *Exptl* | *50 phr N660* | *40 phr N339* |
| M100, MPa | 1.4 | 1.16 | 0.94 | 1.03 |
| Tensile strength, MPa | 18 | 18.3 | 17.1 | 18.4 |
| EB, % | 590 | 630 | 720 | 710 |
| Hardness, IRHD | 54 | 50 | 49 | 45 |
| Crescent tear, N/mm | 24 | 18.6 | 56 | 79 |
| Akron abrasion, vol. loss, $mm^3$ | 111 | 49 | 78 | 38 |
| HBU, °C | 94 | 112 | 77 | 123 |
| Comp set, 1 day/70 °C | 30 | 30 | 26 | 28 |
| Fatigue life at constant 0–100% strain, kcs to failure | 96 | 400 | >650 | >650 |

40 phr of the same black type (N339) as used in the NR/modified-EPDM blend are also shown. It is interesting to note the poorer heat build-up performance of the N339 NR/BR blend in comparison with the NR/modified-EPDM blend compound A. Crescent tear strength, on the other hand, is exceptional for the N339 NR/BR blend, as is abrasion resistance. The fatigue life of the NR/modified-EPDM blend compound A under conditions of 0–100% strain and measured at constant maximum strain, that is with set continuously removed, demonstrates a considerable improvement in performance over the earlier NR/modified-EPDM blends. Improved crack growth behaviour was also observed.

## 15.6 LARGE-SCALE FACTORY MIXING TRIALS

Large-scale factory mixing trials were carried out at the facilities of a medium-sized European tyre manufacturer using the compound A formulation described in Table 15.6. Masterbatch and cross-blended compound were prepared using a computer controlled No. 11 Banbury mixer (270 litre batch size) dumping on to a two-roll mill. It is essential that the optimum operating conditions for the particular mixer used are known both to enable accurate and reproducible temperature profiles of the modified-EPDM masterbatches to be achieved as well as to ensure efficient mixing of the compound. Following minimal working on the mill, the batch was sheeted off, cooled and festooned according to normal factory practice – so demonstrating that the compound could be processed satisfactorily in a factory environment.

A typical NR masterbatch mixing cycle for a No. 11 Banbury mixer is shown in Table 15.8, although this may require modification to suit the particular mixer being used. The main requirement is to achieve good mixing and good black dispersion, as with any sidewall compound. The EPDM mixing cycle, shown in Table 15.9, requires a high rotor speed (60 rpm) initially to get the batch up to the critical reactive mixing

**Table 15.8** NR masterbatch mix cycle for No. 11 Banbury

| *Operation* | *Time (sec)* | *Temperature (°C)* |
|---|---|---|
| Load polymer | 0 | |
| Load powders + ½ black + oil | 30 | |
| Load remainder black | 120 | |
| Ram lift (10 sec duration) | | 105 |
| Dump to temperature | | 130 |

Initial mixer temperature: 70 °C; rotor speed: 30 rpm

**Table 15.9** EPDM masterbatch mix cycle for No. 11 Banbury

| *Operation* | *Time (sec)* | *Temperature*[a] *(°C)* |
|---|---|---|
| Load polymer + powders + black | 0 | |
| Reduce rotor speed and reduce ram pressure | | 145 |
| Dump<br>90 sec after reducing rotor speed | | ~160 |

Initial mixer temperature: 70 °C; rotor speed: initially 60 rpm, reduced to 40 rpm
[a] Indicated mixer temperature (actual batch temperature 5–10 °C higher)

temperature of 150 °C as quickly as possible. The rotor speed is then reduced to 40 rpm to maintain an actual batch temperature of 160 °C ± 5 °C for between 1 and 1.5 mins. Thus, the difference in temperature indicated by the mixer and the actual batch temperature needs to be ascertained accurately. This will depend upon the fill factor employed (0.82 is suggested) and the type of mixer used. For compound A, typical Mooney viscosities (ML (1 + 4) at 100 °C) achieved are as follows: NR masterbatch ~100, EPDM masterbatch 55–70, cross-blend ~60 and the finalized batch ~55, although to some extent this is dictated by the requirements for subsequent extrusion.

## 15.7 COMPOUND EXTRUSION AND TYRE BUILDING

A modern, high performance, low profile tyre – 195 × 65 × R15 V tyre speed rated (240 kph) – was chosen for construction after consultation with a large European tyre company. This size tyre would enable a thorough evaluation of sidewall performance to be made under the most demanding of conditions. The tyre requires extrusion of two components containing the NR/modified-EPDM compound produced in the factory mixing trials, the chafer/sidewall (referred to as the chawall in Figure 15.1) and the treadwing. The chawall is a co-extrusion of the sidewall and the chafer which is a hard NR/BR compound and wraps around the bead in a tyre. The treadwing is a co-extrusion of the tread compound with a small fillet of sidewall compound at each edge designed to give continuity at the critical joint between the tread and the sidewall. Both the chawall and treadwing extrusions have very acute angled edges and are very sensitive to extrusion conditions.

The chawall was co-extruded using twin, cold feed extruders (200 mm and 150 mm) by using a die normally used for production of NR/BR chawall extrudate for the same size tyre. Whilst the quality of the extrudate was deemed acceptable for the purposes of building experimental tyres, it was slightly oversized in width and thickness and had a rather uneven or 'feathered' edge. Satisfactory co-extrusion of the treadwing was achieved using twin 200 mm hot feed extruders, again without any allowances being made for the use of NR/EPDM sidewall compound A.

Tyres were built in two stages by using manually operated tyre building machines. The first stage involved the assembly of the inner liner, casing, beads and chawall. Although poor building tack has often been associated with NR/EPDM blends by others [13], this problem was not encountered in this building trial and confirmed the results of earlier laboratory tests of the building tack of 60:40 NR:EPDM blends. The second stage of building – adding the breakers, restrictor and treadwing – also proceeded without difficulty. Curing of the experimental tyres was carried out under normal factory conditions, which were similar to those used in the laboratory. Thirty-one tyres were built in total with all but one passing the initial quality control check of dimensions, weight and air tightness.

## 15.8 INITIAL TYRE TESTS

Two types of testing were used to assess the tyre's sidewall performance, endurance and high speed testing. For endurance testing, four experimental tyres were used; two tyres were tested on the tyre manufacturer's test rig facilities and two tyres were tested at the MRPRA laboratory. In addition, two identical in specification production tyres were run as controls at the MRPRA facility. The test procedure consisted of five separate legs as shown in Table 15.10. In addition, the tyres were run slightly under-inflated (26 psi) to further increase the severity of the test.

Although the first set of experimental tyres using sidewall compound A were found to pass the first four legs of the endurance test without any

**Table 15.10** Tyre endurance testing programme

| *Leg* | *Duration (h)* | *Load* | |
|---|---|---|---|
| | | *kg* | *% of max* |
| Test speed 80 kph: | | | |
| 1 | 4 | 492 | 80 |
| 2 | 6 | 554 | 90 |
| 3 | 24 | 615 | 100 |
| Test speed 110 kph: | | | |
| 4 | 24 | 830 | 135 |
| 5 | 99 | 923 | 150 |

indications of failure, within the first 30 hours of the final leg all four of the experimental tyres failed. Failure of the experimental tyres was not catastrophic and did not result in deflation. Bulges produced by large cracks in the sidewall, or loose flaps of rubber, in the case of tread separation, were detected by electromechanical sensors fitted to each of the rig's stations. The control tyres (NR/BR sidewall) had shown no sign of failure and would be expected to complete the final leg of the endurance test without any difficulty.

Failure appeared to be of two different types or modes. The first mode of failure was the formation of a series of short diagonal cracks running parallel to each other in the lower third of the sidewall (Figure 15.6). Cracking was preceded by the appearance of an indentation or furrow in the sidewall in a complete ring around the tyre. The indentation was

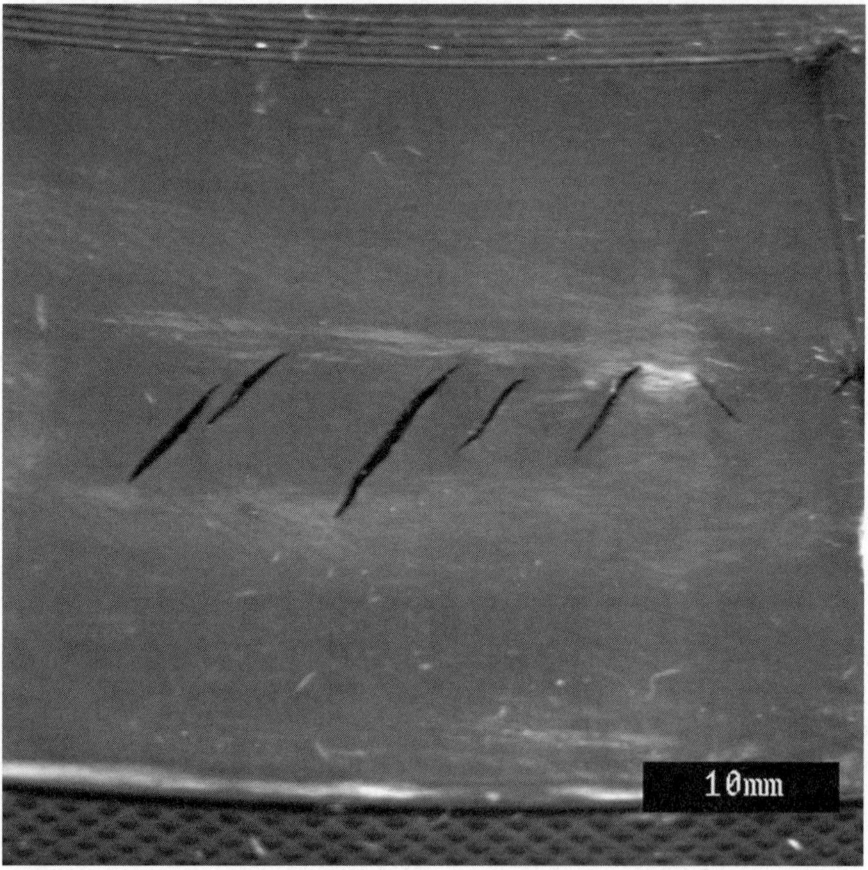


**Figure 15.6** Short diagonal cracks low down on sidewall.

most apparent when the tyre was still hot from testing, although it did not disappear completely on cooling. This indentation was observed in both the experimental and control tyres although cracking only occurred in the experimental tyres. In the final stages of the test, when in parts of the sidewall the short diagonal cracks had formed close together, they would join up to form a circumferential split, typically 10 cm long. It was apparent, however, that these cracks did not originate at the surface of the NR/EPDM sidewall compound, but deep within the tyre sidewall in the chafer/bead region near the ply turn-up of the carcass compound (Figures 15.1 and 15.7).

The second mode of failure appeared as a diagonal split higher up on the sidewall (Figure 15.8), in the upper third, which gradually grew in length and spread into the tread region, ultimately resulting in delami-

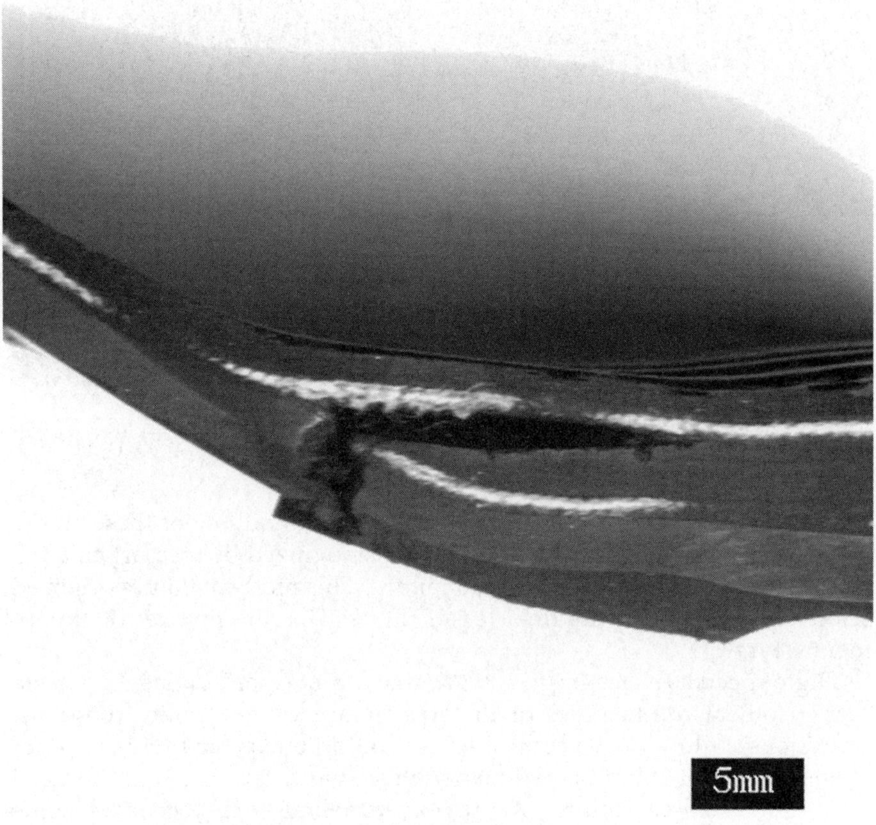


**Figure 15.7** Origin of lower sidewall cracks.

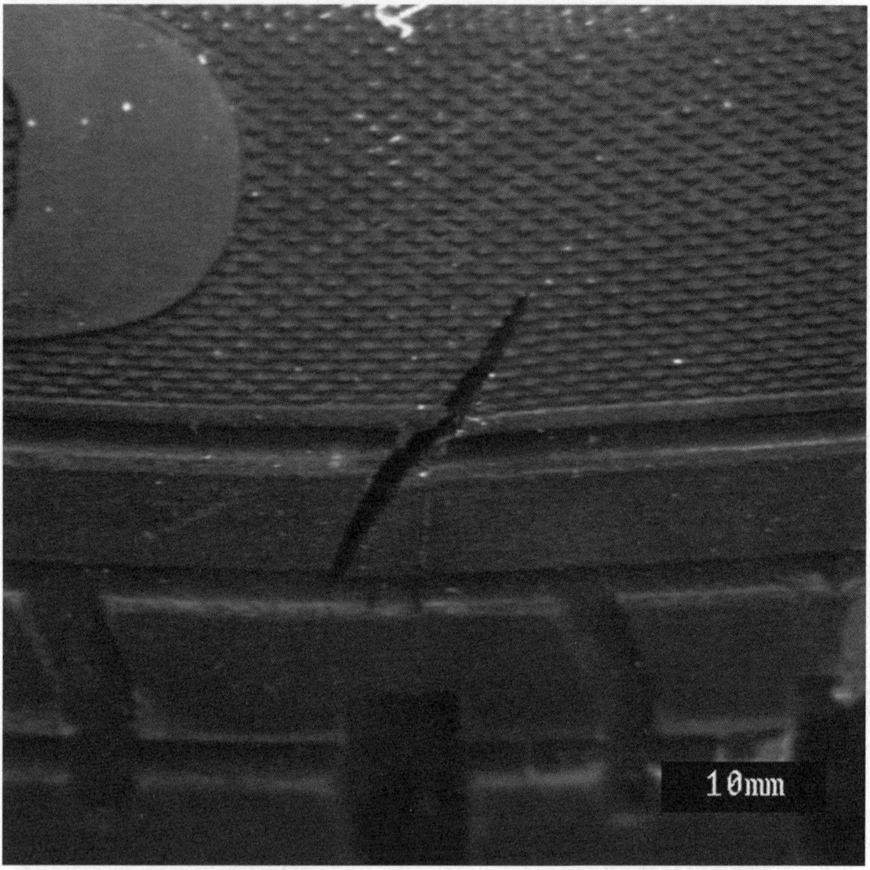


**Figure 15.8** 'Classic' diagonal crack in sidewall/tread region.

nation of the tread. There would be typically only a few of these cracks per tyre, usually less than three. This was recognized by the participating tyre company as a 'classic' sidewall failure. This mode of failure occurred relatively early on in the final leg of the test for the first set of experimental tyres.

High speed tyre testing (115% of maximum speed rating) has been carried out at the facilities of the tyre manufacturer. Under these test conditions, failure of the control tyres would be expected to result after about 2.5 min at 165 mph and be manifested as a separation of the tread from the carcass. Failure of the experimental tyres occurred some 3 minutes into the test. However, it was a matter of concern that failure was in the sidewall region not in the tread, despite the time to failure being in fact longer than that for the control tyre. Inspection of the failed

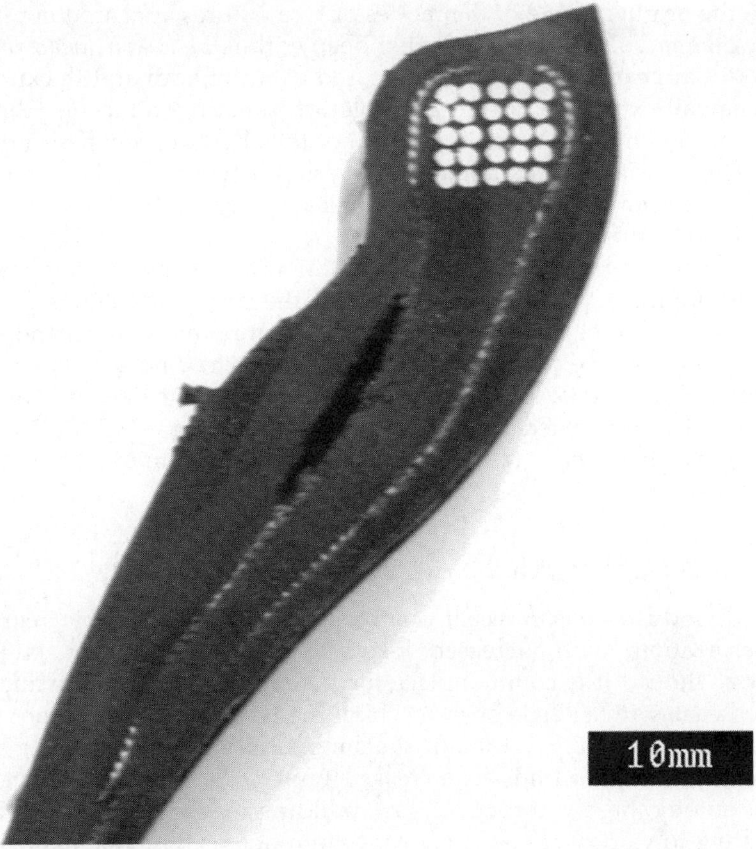


**Figure 15.9** Failure deep within chafer/bead region after high speed testing.

tyre section indicated that the failure was occurring deep in the bead/ chafer region in the thickest region of the sidewall, again not in the sidewall compound itself (Figure 15.9).

## 15.9 ANALYSIS OF INITIAL TYRE TESTING

It is believed that one of the main reasons for premature sidewall failure in the experimental tyres containing NR/modified-EPDM was the unsatisfactory quality of the chawall extrusion. Not only was the chawall extrudate out of specification in terms of dimension, but it was also of inferior quality and would have been unacceptable for a true production run. Relatively minor changes in the sidewall dimensions of a modern high performance, low profile tyre design could be expected to have a critical effect on the behaviour of the tyre under extreme test conditions.

Since the origins of one of the modes of tyre failures appeared not in the sidewall compound itself, but rather deep within the chafer/bead region, it would appear likely that the increased overall width and thickness of the chawall extrudate was responsible for initiating failure in other regions of the tyre. The increased width of the chawall section also meant that there was a reduced amount of overlap between the chafer and the bead filler components which would lead to an increased stress concentration in this region.

It is usual in the development of any new tyre compound that some die development work is required before the compound enters production. As the rheological properties of every compound will depend upon its formulation, the production of a chawall extrudate that lies within specified limits for tyre building will depend upon the design of the die. It would be anticipated that tyres constructed using a NR/EPDM chawall extrudate of the correct dimension would perform to a higher level than those tested.

## 15.10 SECOND FACTORY TYRE BUILDING AND TESTING TRIALS

The tyre industry has in recent years undergone a considerable degree of modernization, with increased levels of automation in the building process. Indeed it is commonplace for a number of semi-automatic machines, each with a single operator, building tyres in two operations to be replaced by one fully automatic machine building tyres in a single operation. Whilst this leads to increased uniformity and higher levels of build quality, highly automated tyre building machines are less accommodating towards out-of-specification components than the manual and semi-automatic building methods they supersede. Thus the extrusion performance of NR/modified-EPDM sidewall compound is increasingly a critical aspect of satisfactory tyre building.

In addition to the problems created by out of specification chawall extrudate, the NR/modified-EPDM sidewall had resulted in a significantly higher modulus than a typical NR/BR sidewall formulation (Tables 15.7 and 15.11). Whilst there are certain merits as discussed above in using a higher modulus sidewall in a tyre, the use of a sidewall compound with a higher modulus than is usual may well be inappropriate for the particular tyre design being used. To address this problem, a second lower modulus formulation sidewall compound (compound B) was prepared as described in Table 15.6. The modulus of this second compound was reduced by lowering the carbon black loading in the NR masterbatch.

Development of a chawall die specifically for the NR/modified-EPDM sidewall compound A was achieved using a further factory mixed batch. Essentially the edges of the chawall were made thicker ($\sim$1 mm thick)

**Table 15.11** Properties of factory mixed NR/EPDM blend compounds

| *Property* | *Compound A* | *Compound B* |
|---|---|---|
| Level of EPDM modification, wt% | 0.138–0.195 | |
| Gel wt% (THF) | 20–24 | |
| M100 | 1.53 | 1.25 |
| M300 | 7.06 | 5.21 |
| TS | 18.9 | 17.9 |
| EB | 550 | 600 |
| Hardness, IRHD | 53 | 50 |
| Crescent tear | 25.7 | 18.8 |
| Heat build-up, °C | 107 | 98 |
| Akron abrasion, vol loss, $mm^3$ | 59 | 54 |
| Compression set, 70 °C/1 day | 32 | 31 |

than the normal NR/BR compound to maintain its integrity as it leaves the die and the die was also made smaller to bring the extrudate within specification. In addition to the minor changes in die design made to accommodate the NR/modified-EPDM blend compound, further changes were incorporated to allow the chawall to be used in tyres built on a fully automatic tyre-building machine.

Extrusion of the chawall using the new die on the equipment described above resulted in a considerable improvement in extrudate quality. The lower modulus compound B gave better extrusion performance, being comparable in this respect to a normal NR/BR sidewall compound. Both NR/modified-EPDM sidewall compounds gave chawall extrudate that was acceptable for automated tyre building. Sixteen tyres were built with each chawall compound using an automatic building machine. The overall construction of the second two sets of tyres was seen as being superior to the first set. The internal transitions between the various tyre components were much neater and the profile of the entire tyre sidewall was much smoother than before.

Endurance testing of a pair of tyres built with each of the two different chawall compound tyres was carried out at the MRPRA facilities in the same manner as described above. High speed testing was not performed. The tyres successfully completed the first four stages of the endurance test, but failure again occurred in the final leg of the test. Unfortunately a mechanical failure of the rig early in the final leg of the test meant that only one of the higher modulus sidewall compound (compound A) tyres was fully tested. The high degree of reproducibility shown in the previous endurance test did indicate that similar behaviour from a second tyre built with compound A could be expected with some confidence.

As before, two modes of failure were observed. The first mode was a long circumferential split occurring low down on the sidewall. The second mode was a diagonal crack higher up on the sidewall leading into the tread region. The compound A tyre showed only the first mode of failure, while the lower modulus compound B tyres showed the first mode of failure in one tyre, and the second mode of failure in the second tyre. There was no evidence in either the compound A or B tyres of the annular sidewall depression or furrow that was seen in the first set of tyres. The time to failure for the first mode of failure was at about 24 hrs which was slightly longer than for the first set of experimental tyres. The time to failure for the second mode of failure was considerably longer than before at 56 hrs. The tyres were not tested to complete destruction, but as before to a point when it was considered that complete failure was imminent.

The results of the second tyre testing trial, although disappointing, were seen as a considerable improvement over the rig tests with the first set of experimental tyres, especially the progression well into the second half of the final leg of the endurance test for the lower modulus compound B tyre. Although a number of design changes had been incorporated in the chawall of the second set of tyres, essentially none of these changes were made specifically on the basis of the NR/modified-EPDM blend compounds having a higher modulus than a typical NR/BR sidewall compound.

As before, the first mode of failure was seen very much as a tyre design problem, as the fault occurred in a region of the tyre which would have had very little direct contribution from the sidewall compound. It was observed that, as was the case with the first set of tyres, some degree of coincidence of the end of the bead filler and the ply turn-up was present which would result in an increased stress concentration in this relatively vulnerable region. Greater consideration of the difference in modulus of the NR/EPDM compound, compared with NR/BR in the design of the tyre overall, would be expected to permit shifting of the area of stress concentration away from the critical ply turn-up region. The observation for the one tyre that was able to overcome the first mode of failure suggests that the changes which were introduced for the second series of tyres did give considerable benefits in terms of the endurance behaviour of the sidewall compound.

The reoccurrence of the second mode of failure was seen as more of a problem associated with the NR/EPDM sidewall compound itself. There would be little scope for specific design changes to improve the performance of the tyre in a region so close to the tread. However, the margin between failure of the lower modulus sidewall compound and it completing the endurance test is relatively small, and minor changes may be sufficient. It was unfortunate that the higher modulus formulation tyre did not progress to a stage of the test where resistance to the second mode of failure could be assessed. Previous crack growth testing indicated that the higher modulus formulation is to be preferred.

## 15.11 OZONE AND ATMOSPHERIC WEATHERING TESTS

Exposure to 50 ppb ozone under static conditions for seven days failed to produce any cracks or surface discolouration in the NR/modified-EPDM sidewall of the compound A experimental tyre. Under the same test conditions the control tyre, although showing no cracks, did exhibit a slight surface discolouration. Exposure of freshly washed experimental and control tyres to atmospheric weathering for three months (July–September 1995) resulted in a considerable amount of blooming and discolouration on the exposed sidewall of the control tyre. The experimental tyre suffered no such effects, being similar in appearance to when it was first exposed to the weathering conditions.

## 15.12 CONCLUSION

An experimental design approach, incorporating the 'reactive mixing' technology for modification of EPDM with sulphur donors, has been used to optimize black and oil loadings in a 60:40 NR:EPDM blend to give properties matching a typical NR/BR sidewall compound in nearly all respects. The best range of properties, in which crack growth resistance is given precedence, is achieved by a compound having a relatively high level of N339 black in the NR phase, a very low level of N339 black in the EPDM, and a low to medium level of paraffinic oil in the blend overall. A proportion of high molecular weight oil extended EPDM is beneficial for enhancing the effectiveness of the modification in improving compression set, fatigue resistance and crack growth properties.

Mixing of such NR/EPDM blends has been successfully demonstrated with a variety of internal mixers up to full factory scale using a 270 litre No 11 Banbury mixer and factory processing equipment. A successful factory tyre-building trial has demonstrated the utility of NR/modified-EPDM sidewall compound at all stages of tyre manufacture.

Premature failure of the experimental tyres in the latter stages of endurance and high speed rig testing demonstrated that direct substitution by a NR/modified-EPDM sidewall compound for an NR/BR sidewall compound in an existing tyre design is inappropriate. Allowances for differences in extrusion and property characteristics of these two different types of compound need to be made. Subsequent changes in die design coupled with minor changes in the overall tyre design to enable tyres to be built on an automatic building machine gave a significant improvement in the quality of the tyre build. The results of a second endurance tyre trial were sufficiently encouraging to suggest that if further changes were incorporated into the overall design of the tyre, a NR/modified-EPDM sidewall should be successful.

## 15.13 REFERENCES


1 Marwede, G.W., Stollfuß, B. and Sumner, A.J. (1993) Current status of tyre elastomers in Europe, *Kautschuk Gummi Kunststoffe*, **46**(5), 380–388.

2 Fusco, J.V. and Young, D.G. (1990) Isobutylene-based polymers in tires – status and future trends. *ACS Rubber Division Meeting*, Washington, DC, USA, October 9–12.

3 Morrissey, R.T. (1971) Halogenation of ethylene propylene diene rubbers. *Rubber Chemistry and Technology*, **44**(4), 1025–1042.

4 Baranwal, K.C. and Son, P.N. (1974) Co-curing of EPDM and diene rubbers by grafting accelerators onto EPDM. *Rubber Chemistry and Technology*, **47**(1), 88–99.

5 Mastromatteo, R.P., Mitchell, J.M. and Brett, Jr., T.J. (1971) New accelerators for blends of EPDM. *Rubber Chemistry and Technology*, **44**(4), 1065–1079.

6 Shell Chemical Company (1965) *Technical Bulletin SCR*: 65–160, August.

7 von Hellens, C.W. (1989) EPDM in black sidewalls. Paper No. 40, ACS Rubber Division Meeting, Detroit, Michigan, USA, October 17–20.

8 Tisler, A. (1992) A review of sulphenamide type acceleration systems for improved environmental desirability and maintenance of compounding flexibility. Paper No. 88, ACS Rubber Division Meeting, Nashville, Tennessee, USA, November 3–8.

9 von Hellens, C.W. (1992) Effect of EPDM characteristics on cure state and co-vulcanization. Paper No. 23, ACS Rubber Division Meeting, Nashville, Tennessee, USA, November 3–8.

10 Brown, P.S. and Tinker, A.J. (1996) The effect of modifying EPDM in the crosslink distribution of NR/EPDM blends. *Journal of Natural Rubber Research*, submitted for publication.

11 Freakley, P.K. (1985) *Rubber Processing and Product Organisation.*, Plenum Press, New York.

12 Davies, V.L. (1954) *Design and Analysis of Experiments*. Oliver & Boyd, London.

13 von Hellens, C.W., Edwards, D.C. and Lobos, Z.J. (1990) Applications of EPDM in tire black sidewalls. *Rubber and Plastics News*, 61.

14 Sumner, A.J.M., Kelbch, S.A. and Eisele, U.G. (1994) Crack growth performance of tire compounds. Paper No. 18, ACS Rubber Division Meeting, Pittsburgh, Pennsylvania, USA, October 11–14.

15 Eisele, U., Kelbch, S. and Engels, H.W. (1992) The tear analyser – A new tool for quantitative measurements of the dynamic crack growth of elastomers. *Kautschuk Gummi Kunststoffe*, **45**(12), 1064–1069.

16 Lake, G.J. and Thomas, A.G. (1967) Physics of failure in rubber. *Kautschuk Gummi Kunststoffe*, **20**(2), 211–217.

# 16

# Compounding NR/EPDM blends for light-coloured applications

*Stuart Cook* and Rose Marie Escolar†*
**Tun Abdul Razak Research Centre, MRPRA, UK;*
*†Power Synthetic Rubber Manufacturing Corp., Manila, Philippines*

## 16.1 INTRODUCTION

The work described in this chapter, although targeted in particular at gaskets for washing machines and washer/driers, is believed to be applicable to non black-filled applications in general, as the method of manufacture for washing machine gaskets, like many other products for use in the domestic appliance market, is by the process of injection-moulding.

Injection-moulding, although originally developed and perhaps more widely known as a processing method for plastics [1], has over the past 35 years become an increasingly important method for the fabrication of rubber products [2, 3]. It is a high speed process which lends itself well to automation. Injection-moulding as a process can be subdivided into a number of separate steps: automatic feeding, heating and plasticizing of a compound followed by injection into a heated mould where vulcanization takes place. Ideally, the injection process heats the compound as close as is practical to the vulcanization temperature by forcing a metered quantity of rubber at high pressure through a restricting nozzle, runner system and gates into the mould cavity. Thus, in conjunction with high cure temperatures, cure times can be significantly reduced in comparison with compression and transfer moulding. In addition, the tolerances of the injection moulded product can be reduced since injec-

*Blends of Natural Rubber*. Edited by Andrew J. Tinker and Kevin P. Jones.
Published in 1998 by Chapman & Hall, London. ISBN 0 412 819406

tion takes place into a closed mould, and trimming and inspection costs can be reduced by the elimination of flash and reduced reject rates.

Washing machine gaskets used to be based on natural rubber (NR) compounds [4] but, due to their poor resistance to ozone and the action of heat and detergents, ethylene-propylene-diene monomer (EPDM) has replaced NR in this application. EPDM compounds [5], however, rely on expensive reinforcing silica fillers and coupling agents to achieve acceptable strength properties. EPDM compounds also tend to have poor hot tear resistance which can lead to problems with demoulding in the manufacturing process. Compounds based on EPDM also have a tendency for oil exudation which may lead to bacterial attack.

Compounds based on blends of NR and EPDM would be anticipated to provide an acceptable compromise between the properties of both NR and EPDM in this application area. The NR phase should provide good physical properties without the necessity for highly reinforcing fillers and expensive coupling agents, whereas the EPDM phase should provide good heat ageing and ozone resistance, without the need for antidegradants. Attention has centred on a 60:40 NR/EPDM polymer ratio which offers protection against ozone attack without the need for conventional antiozonants [6].

Modification of EPDM with the sulphur donors *bis*-alkylphenol disulphide (BAPD) and dithiodicaprolactam (DTDC) prior to cross-blending with NR has been shown (see Chapter 14) to be effective in both unfilled and black-filled blends for reducing the phase size of the blends and increasing the crosslink density of the EPDM phase. Both of these effects have been shown to lead to improvements in general physical properties. A similar general approach to that described for black-filled NR/EPDM blends (Chapters 15 and 17) has been employed in work on white-filled NR/EPDM blends.

## 16.2 GENERAL EXPERIMENTAL PROCEDURES

The preparation of white-filled NR/modified-EPDM compounds followed essentially the same procedure as that described in Chapter 14; separate NR and EPDM masterbatches are prepared, to allow modification of the EPDM, then cross-blended to give a 60:40 NR:EPDM blend. The NR and EPDM grades used were SMR L and generally Polysar 585, a 'highly' unsaturated (ENB ter-monomer) EPDM grade. Where other grades of EPDM were used, the commercial name is indicated.

Small-scale masterbatch and cross-blend mixing was performed in a Brabender PL2000E fitted with a 350S mixing head and Banbury style rotors (300 ml capacity, 0.7–0.75 fill factor), or a 00C Banbury (3 litre capacity, 0.75 fill factor) internal mixer. Larger scale mixing has been carried out in a Farrel F50 Banbury mixer fitted with Synchronous

Technology (ST) rotors (47 litre capacity, 0.72 fill factor). In all cases the most important step is to achieve a satisfactory level of modification in the EPDM masterbatch. For the Farrel F50, masterbatches were typically mixed at a high rotor speed (70 rpm) to an indicated temperature of 165 °C before dumping; the temperature of the dumped batch being ~175 °C. The batch ingredients are added early on in the mix cycle after which the ram remains down for the remainder of the mix cycle. This enables the EPDM masterbatch to reach, and then maintain a temperature of ~160° to ~175 °C for a period in the region of 1.5–2 minutes, to allow modification to occur. The total mixing time of the EPDM masterbatches is about 4–5 minutes. Cross-blending is typically achieved using a two-minute mix cycle at a medium rotor speed. Finalizing was carried out either on a two-roll mill or in an internal mixer at low rotor speed.

Vulcanized test pieces (buttons and sheet) were prepared mainly by injection moulding at 180 °C using a B43K REP injection moulding machine. Cure times were determined by injection moulding compression set buttons at various cure times based on the $t_{95}$ time derived from a Monsanto MDR 2000 rheometer trace. Compression set testing of the buttons was performed to determine the optimum cure time, i.e. the cure time giving the lowest compression set value. Cure conditions cannot reasonably be compared between a rheometer and an injection moulding machine where the sample size, heat transfer and degree of temperature control are significantly different.

## 16.3 PREPARATION OF BLEND COMPOUNDS

To demonstrate the feasibility of using NR/EPDM blends for the washing machine gasket application, a series of three 60:40 NR:EPDM blends were prepared and their properties compared against NR and EPDM control compound formulations. The formulations of all of these compounds are depicted in Tables 16.1 to 16.3. The NR control formulation, on which the blend compounds were based, had been shown in earlier studies [4] to give a good balance of physical properties. The EPDM control compound was typical of a commercial EPDM washing machine gasket formulation [5]. One of the blend compounds contained unmodified EPDM to give a measure of the effectiveness of the EPDM modification procedure in improving physical properties. In addition, the antidegradants Wingstay L (1 phr) and TMQ (3 phr) were added with the curatives to provide a reasonable degree of protection against oxidation to the NR control compound and to one of the NR/modified-EPDM blends.

As described in Chapter 14, a separate masterbatch and cross-blending procedure has been adopted to facilitate modification of only the

**Table 16.1** NR washing machine gasket formulation [4]

| *Ingredient* | *phr* |
|---|---|
| SMR L | 100 |
| Winnofil S[a] | 40 |
| Titanium dioxide | 3 |
| Strukpar 2280 | 15 |
| Struktol WB 212 | 2 |
| Zinc oxide | 5 |
| Stearic acid | 1 |
| Sulphur | 0.6 |
| TBBS | 1.0 |
| Cure Rite 18[b] | 1.7 |

[a] Coated calcium carbonate
[b] N-oxydiethylene-thiocarbamy-N′-oxydiethylene sulphenamide

**Table 16.2** Typical commercial EPDM washing machine gasket formulation [5]

| *Ingredient* | *phr* |
|---|---|
| Vistalon 3666 | 125 |
| Vistalon 8600 | 30 |
| Polestar 200R[a] | 60 |
| Titanium dioxide | 10 |
| Ultrasil VN3 | 30 |
| Durosil | 30 |
| Enerpar 13 | 40 |
| Zinc oxide | 5 |
| Stearic acid | 1 |
| PEG 4000 | 2 |
| Polyvest 25[b] | 2 |
| Flectol Pastilles | 0.5 |
| Sulphur | 0.5 |
| MBTS | 1.5 |
| DPTH | 1.5 |
| Rhenocure ZAT[c] | 1.4 |
| Vulkacit I[d] | 1.0 |

[a] Clay
[b] Silane coupling agent
[c] Zinc amine dithiophosphate complex
[d] Dimethyl diphenyl thiuram disulphide

**Table 16.3** NR control and blends formulation

| | *NR control* | *Unmodified blend* | *Modified blend* | *Modified blend* |
|---|---|---|---|---|
| SMR L | 100 | | | |
| 60:40 NR:EPDM | | 100 | 100 | 100 |
| BAPD | – | – | 0.8 | 0.8 |
| DTDC | – | – | 0.4 | 0.4 |
| Wingstay L | 1 | – | – | 1 |
| TMQ | 0.5 | – | – | 0.5 |

All contain 40 phr Winnofil S, 3 phr titanium dioxide, 15 phr Strukpar 2280, 2 phr Struktol WB 212, 5 phr zinc oxide, 1 phr stearic acid, 0.6 phr sulphur, 1 phr TBBS, 1.7 phr Cure Rite 18

EPDM phase of the blend. Modification of EPDM (Polysar 585) was achieved with 0.8 phr BAPD and 0.4 phr DTDC. This combination of sulphur donors has been found to give a good overall balance of properties, especially for high temperature vulcanization (Chapter 14). Where a different modification system has been used this is noted.

The modification procedure for white-filled EPDM masterbatches is somewhat more flexible than for the black-filled masterbatches described in Chapters 15 and 17, as there is little interaction between the modified EPDM and white filler. This leads to a rather slower rate of temperature increase in the mixing chamber and a very much lower mix viscosity than for a black-filled mix. It is, however, possible to achieve relatively high levels of efficiency of EPDM modification in a white-filled compound (up to ~0.5 wt%), without the batch becoming unprocessable in the manner encountered when a black-filled mix is allowed to reach too high a temperature for prolonged periods. It is thought that less modifier is absorbed during mixing by non-interactive fillers such as calcium carbonate than by carbon black, thus making more sulphur donor available for binding to the EPDM. In addition, more EPDM-bound modifier may remain available for crosslink formation than in a black-filled blend, as a proportion of the polymer-bound modifier could be rendered 'unreactive' with regard to crosslinking reactions by the high degree of binding of the modified EPDM to carbon black that takes place.

The processing characteristics of the NR/EPDM blends on the open mill are good; they are tacky and cohesive, and are generally easier to process than the pure EPDM compounds. In terms of cure behaviour, the modified EPDM blends have similar cure times to the NR compound, whereas the unmodified EPDM blends have considerably longer cure times. Each compound was injection moulded at a cure temperature of

180 °C as described above to prepare a range of test pieces. A typical cure time was in the order of 2.5 minutes. Ozone, weathering and detergent-resistance tests were performed in addition to the usual unaged and hot air ageing tests.

## 16.4 COMPARISON OF PHYSICAL PROPERTIES

Comparing the unaged physical properties of each compound (Table 16.4) it is seen that the modification procedure does improve the general blend properties although the tensile strength, hot tear resistance and compression set performances of the blends all fall short of the performance offered by either the NR or EPDM controls. It should be noted that the level of EPDM modification achieved in the blends used in this study was rather low at 0.11 wt% as determined by the GPC/UV method described in Chapter 14; a level of double or even triple this value would normally be regarded as satisfactory. An improved level of EPDM modification would be expected to bring further improvements to the blend properties. The addition of antioxidant to the modified blend does improve tensile strength and elongation at break marginally, but rather unexpectedly reduces hot tear strength to a very low level. During injection moulding, however, the presence of antioxidant does minimize the tendency for flash oxidation and thus facilitates flash removal and enables shorter cycle times.

The retention of properties after hot air ageing (72 hours at 100 °C) is shown in Table 16.5. Here it is seen that the blend compounds generally compare much more favourably with the EPDM control compound and

**Table 16.4** Unaged physical properties

| | *NR control* | *Unmodified blend (no A/O)* | *Modified blend (no A/O)* | *Modified blend (with A/O)* | *EPDM control* |
|---|---|---|---|---|---|
| $t_{95}$ at 180 °C, min:sec | 3:12 | 4:22 | 3:07 | 3:00 | 3:41 |
| M100, MPa | 0.49 | 0.66 | 0.73 | 0.69 | 0.93 |
| TS, MPa | 19.3 | 7.99 | 8.58 | 9.37 | 11.6 |
| EB, % | 860 | 745 | 710 | 750 | 810 |
| Hardness, IRHD | 29 | 36 | 40 | 39 | 47 |
| Hot trouser tear (100 °C), N/mm | 4 | 2.5 | 2.8 | 1.9 | 5.9 |
| Compression set (1 day/70 °C), % | 20 | 36 | 36 | 37 | 25 |

**Table 16.5** Percentage changes in properties on ageing in air for 72 h at 100 °C

| | *NR control* | *Unmodified blend (no A/O)* | *Modified blend (no A/O)* | *Modified blend (with A/O)* | *EPDM control* |
|---|---|---|---|---|---|
| TS, MPa | −45 | −26 | −26 | −18 | +3 |
| EB, % | −17 | −8 | −8 | −13 | −9 |
| Hardness, IRHD | +7 | −15 | −8 | +3 | +18 |
| Compression set (1 day/70 °C), % | −25 | +6 | −3 | −8 | −28 |

**Table 16.6** Percentage changes in properties on ageing in hot detergent ('Original formula' non-biological Persil automatic powder, 10 g/litre, 72 h, 95 °C)

| | *NR control* | *Unmodified blend (no A/O)* | *Modified blend (no A/O)* | *Modified blend (with A/O)* | *EPDM control* |
|---|---|---|---|---|---|
| TS | −5 | +7 | +2 | +4 | −11 |
| Hardness | −2 | −8 | −10 | −6 | −12 |
| Compression set | −35 | −6 | −8 | −14 | −48 |
| Volume change, % | +3.7 | +3 | +2.4 | +2.9 | +12.4 |

have a considerably better property retention performance than the NR control compound. Modified blends continued to out-perform the unmodified blend, and the addition of antioxidants generally improves property retention, especially for tensile strength and hardness.

Hot detergent ageing and volume swelling tests (Table 16.6), to simulate service conditions typically encountered by washing machine gaskets, showed the blends to have generally better resistance than either the NR or EPDM controls. The modified and the unmodified blends performed similarly, with antioxidants having little overall impact. The particularly high uptake of water in the case of the EPDM control compound is probably a result of the presence of silica in the compound and would be expected to be lower in a truly commercial formulation.

## 16.5 FORMULATION DEVELOPMENT

The general effectiveness of the EPDM modification in improving the basic physical properties of NR/EPDM blends has been shown above. This is considered to have been a result of the improved crosslink density

**Table 16.7** Crosslink densities in filled NR/EPDM blends ($t_{95}$ at 180 °C)

| *Vulcanizate* | *Crosslink density, mol/m*$^3$ | |
|---|---|---|
| | *NR* | *EPDM* |
| NR control | 53 | – |
| NR/EPDM | 71 | 9 |
| NR/modified-EPDM | 80 | 15.5 |

EPDM modification level: 0.452 wt% using BAPD 1.0 phr). Cure system: 0.6 phr sulphur, 1 phr TBBS, 1.7 phr Cure Rite 18

achieved in the EPDM phase as a result of modification with sulphur donors; the importance of improving the crosslink density of the EPDM phase of a blend with NR has been emphasized in Chapter 14. It can be seen in Table 16.7 that modification of EPDM (Polysar 585) using the sulphur donor BAPD results in an increase in the EPDM crosslink density, as measured by swollen-state $^{13}$C NMR [7], of over 50%, although it remains at a very low level in comparison with the crosslink density in the NR phase. A smaller but still significant increase in crosslink density is also seen in the NR phase of the blend; this is most probably a result of utilization by the NR of unreacted sulphur donor or active fragments of the sulphur donor made available by the EPDM phase.

Higher levels of crosslinking in the EPDM phase might be expected to improve the physical properties of blends, especially the compression set performance. This may be achieved in a number of ways, one of which would be to make better use of the existing cure system. This can be achieved by using a higher molecular weight EPDM which will give a higher physical crosslink density for a given curative level.

The effects of part replacement of the Polysar 585 by Polysar 5875, a high molecular weight, highly oil-extended grade of EPDM can with a relatively high level of unsaturation (ENB) be seen in Table 16.8. In terms of improvement of compression set, a higher proportion of oil-extended EPDM and lower filler level is favourable. However, the use of Polysar 5875 results in a reduced modification efficiency, very probably caused by a lower temperature rise during the modification step resulting from the high level of oil extension (100%) and hence low compound viscosity. The hot tear resistance of compounds containing Polysar 5875 is also poor, resulting in tearing of the test pieces during demoulding. This, as discussed below, can be attributed to the increased oil level in the blend as a consequence of using this grade of EPDM. The suitability of different grades of EPDM for injection moulding applications has not been investigated fully, although it is realized that the processing behaviour of

**Table 16.8** Effect of EPDM type and filler loading on properties

| | | | | | |
|---|---|---|---|---|---|
| Polysar 585 | 50 | 50 | 70 | 50 | 100 |
| Polysar 5875 | 100 | 100 | 60 | 100 | – |
| Winnofil S | 40 | 60 | 40 | 80 | 40 |
| EPDM modification, wt% | 0.065 | 0.1 | 0.127 | 0.161 | 0.185 |
| Compression set (70 °C/22h, %) | 32 | 33 | 34 | n/a | 37[a] |
| Ease of demoulding | Poor | Poor | Poor | Poor | Good |

[a] Denotes compression set buttons compression moulded at 180 °C rather than injection moulded
All contain 3 phr titanium dioxide, 3 phr Strukpar 2280 oil, 5 phr zinc oxide, 1phr stearic acid, and 0.8 phr BAPD/0.4 phr DTDC as modifiers. Cure system: 0.6 phr sulphur, 2.0 phr TBBS, 1.0 phr TMTD, 0.25 phr Tetrone A, 0.5 phr TBTD, 1.0 phr ZDBC

the blends is strongly dependent upon the type of EPDM grade selected. To gain the greatest benefit from modification with sulphur donors, the EPDM grade should have a high level of ENB unsaturation (Chapter 14).

Increased curative levels in the blend should also be expected to increase the level of crosslinking in the EPDM phase and to reduce compression set. A 15% increase in the overall level of curatives produces a slight improvement in compression set, although less than the effect of replacing half of the Polysar 585 by Polysar 5875. A higher overall curative level, however, produces an excessively high level of crosslinking in the NR phase which reduces strength properties.

Changes to the filler and oil loading, as shown in Table 16.9, can have a significant effect upon the properties of the blend vulcanizate.

**Table 16.9** Effect of oil and filler loading on physical properties

| *Oil loading, phr* | | *Property* | *Filler loading, phr* | |
|---|---|---|---|---|
| *15*[a] | *5*[a] | | *40*[b] | *85*[b] |
| 8.64 | 10.2 | Tensile strength, MPa | 11.3 | 9.85 |
| 655 | 650 | Elongation at break, % | 685 | 710 |
| 39 | 45 | Hardness, IRHD | 45 | 52 |
| 2.1 | 4.0 | Hot trouser tear (100 °C), N/mm | 5.9 | 9.4 |
| 35 | 35 | Compression set at 70 °C/22 h, % | 33 | 36 |

[a] EPDM modified with 1.0 phr BAPD
[b] EPDM modified with 0.8 phr BAPD/0.4 phr DTDC
All contain 3 phr titanium dioxide, 5 phr zinc oxide, 1 phr stearic acid, and 0.8 phr BAPD/0.4 phr DTDC as modifiers. Cure system: 0.6 phr sulphur, 2.0 hr TBBS, 1.0 phr TMTD, 0.25 phr Tetrone A, 0.5 phr TBTD, 1.0 phr ZDBC

Reduction of the oil level to 5 phr, in comparison with the compound formulation shown in Table 16.3, is beneficial in nearly all respects, leading to improved tensile strength and hot tear resistance. Increasing the filler loading from 40 to 85 phr improves hot tear resistance significantly and increases hardness, but produces a compound with lower tensile strength and inferior compression set.

## 16.6 WEATHERING AND OZONE TESTING

Accelerated daylight ageing tests were conducted to assess the performance of the light-coloured blends in more general outdoor applications. Samples were exposed for 250 hours (500 hours' total running time) to a high intensity light source in a Xenotest 150S Weathering Machine in which a continuously air-cooled test chamber is illuminated by a Xenon lamp fitted with IR and UV filters to give a high intensity daylight spectrum light. The intensity of light produced (1053 $Wm^{-2}$) represents, depending upon latitude [8], between three and five times the average incident solar radiation under cloudless conditions. The presence of cloud cover reduces the solar radiation on the ground by about 70% [9], thus Xenotest ageing, amounting to 250 hours' exposure, represents around six months of 'real time' exposure.

The Xenotest sample holders shield part of the sample from exposure to any light, therefore to ensure a comparable heat history for the test pieces, samples were taken from both exposed and unexposed areas and tensile properties compared. Table 16.10 indicates that retention of tensile properties after exposure to high intensity daylight ageing conditions were generally better for the blends than for either the NR or EPDM control compounds; the latter performed particularly poorly. All test pieces were discoloured to some extent and showed surface cracks after

**Table 16.10** Percentage change in properties after high intensity 'Daylight' exposure (Xenotest, Level 1, 500 h total running time, 250 h 'Daylight' exposure)

| | *NR control* | *Unmodified blend (no A/O)* | *Modified blend (no A/O)* | *Modified blend (with A/O)* | *EPDM control* |
|---|---|---|---|---|---|
| M100 | +54 | +25 | +22 | +34 | +77 |
| Tensile strength | −35 | −46 | −50 | −50 | −65 |
| Elongation | −17 | −13 | −17 | −20 | −50 |

Samples have the same heat history

exposure, but the NR control was by far the worst in this respect. The unmodified blend showed slightly better performance than the modified blends and the presence of antioxidants slightly reduced the resistance of the NR/modified-EPDM blend to light ageing and led to some surface discolouration.

The ozone resistance of NR/EPDM blends is far better than that of the NR control. After exposure to 50 pphm ozone (40 °C, 100% strain) for 72 hours, the NR control already showed severe surface cracking and discolouration. The NR/EPDM blends, on the other hand, showed no signs of cracking at all. Phase morphology studies of white-filled blends, using STEM, have shown the modification of EPDM with sulphur donors to be effective in reducing the phase size of EPDM from approximately 2 μm to 1 μm. This would be expected to benefit ozone resistance still further.

## 16.7 CONCLUSION

The ability to compound and produce NR/modified-EPDM blends to match many of the properties of a typical EPDM washing machine gasket formulation has been demonstrated. This has been achieved without the use of highly reinforcing fillers or expensive coupling agents. Moreover, it was shown that the NR/modified-EPDM blends give good ozone resistance, and generally better aged property retention after exposure to various media, than the EPDM compound.

The versatility of the EPDM modification procedure using commercially available sulphur donors has been successfully demonstrated for white-filled compounds. The modification is effective in a variety of commercially available EPDM grades having differing molecular weights, degrees of unsaturation, ethylene:propylene ratio, and oil extension. In all cases the effect of the modification is to improve the general physical properties of a blend with NR by increasing the physical crosslink density of the EPDM phase. In addition, the EPDM modification technique has been shown to be effective in large-scale mixing trials (Farrel F50 Banbury, 47 litre batch size). White-filled NR/modified-EPDM blends have also been shown to be suitable for injection moulding using conditions typical of those used to process NR. These tend to be at lower temperatures than for EPDM.

The washing machine gasket formulation has yet to be optimized in terms of cure system and filler and oil loading; further improvements in compression set performance are desirable. The use of alternative fillers and changes to the cure system might be expected to go some way to producing a compound with more acceptable stability at high temperatures. Despite the rather limited nature of the investigation, it is apparent that the general principles of EPDM modification are successful.

## 16.8 REFERENCES


1 Crawford, R.J. (1987) *Plastics Engineering*. Pergamon Press, Oxford.

2 Wheelans, M.A. (1971) Injection moulding with natural rubber. *Rubber Chemistry and Technology*, **44**(3), 620–641.

3 Paris, W.W., Dillhoefer, J.R. and Woods, W.C. (1982) Accelerator systems for injection-moulding. *Rubber Chemistry and Technology*, **55**(2), 494–512.

4 MRPRA (1984) *The Natural Rubber Formulary and Property Index*.

5 Personal communication, Exxon Chemical Ltd, Fareham, Hampshire, UK.

6 Brown, P.S. and Tinker, A.J. (1996) The effect of modifying EPDM in the crosslink distribution of NR/EPDM blends. *Journal of Natural Rubber Research*, submitted for publication.

7 Loadman, M.J.R. and Tinker, A.J. (1989) The application of swollen-state CW-1H NMR spectroscopy to the estimation of the extent of crosslinking in vulcanized polymer blends, *Rubber Chemistry and Technology*, **62**, 234–245.

8 Weast, R.C. (ed.) (1980) *The Handbook of Chemistry and Physics*, 61st edition. CRC Press, Boca Raton, Florida, USA.

9 Personal communication, Meteorological Office, Bracknell, Berkshire, UK.

# 17

# NR/EPDM blends for extruded profile weatherstrip

*Mohamed Aris bin Ahmad**, *Ridha Arizal*†
*and Stuart Cook*§
**Rubber Research Institute of Malaysia;* †*Bogor Research Station for Rubber Technology, Indonesia;* §*Tun Abdul Razak Research Centre, MRPRA, UK*

## 17.1 INTRODUCTION

The seals for automotive windows, doors, lights, etc. may all be considered as weatherstrip or weatherseals. The production of weather strip usually involves the extrusion of a complex shaped profile with precisely defined dimensions (see Figure 17.1). Rubber has for many years been used for sealing purposes in a vehicle; it is flexible and resilient and can be extruded relatively easily. In the early 1970s, automotive weatherstrip was made from a single polymer rubber, mainly natural rubber (NR) which has excellent resilience, but relatively poor weathering resistance. However, other elastomers including styrene-butadiene rubber (SBR), polychloroprene (CR) and ethylene-propylene-diene monomer rubber (EPDM) were also used [1]. Towards the end of the 1970s EPDM had become dominant primarily because of its superior weathering and ozone resistance coupled with its high filler acceptance and wide processing latitude. Analysis [2] of a recent, medium-sized, high sales volume vehicle has shown that more than 66% of the non-tyre rubber used in the car is based on EPDM of which some 75% is EPDM weatherstrip of some form. This compares with about 12% of the non-tyre rubber in a car being NR.

*Blends of Natural Rubber*. Edited by Andrew J. Tinker and Kevin P. Jones.
Published in 1998 by Chapman & Hall, London. ISBN 0 412 819406

**Figure 17.1** Photograph of an extruded weatherstrip.

Modern weatherstrip for automotive door and boot seals commonly consists of solid or co-extruded solid and cellular profiles whilst window seal profiles can contain a metal insert for added stiffness or be multiple co-extrusions of different hardness compounds [3]. Rapid curing of the extruded profile is achieved at high temperature by a continuous vulcanization procedure, such as microwave, hot air line, liquid cure medium (LCM), or the fluidized bed method. There are a number of important considerations to be made in choosing a compound for use as weatherstrip. In its uncured state, the extruded profile must possess high green strength to prevent collapse and so maintain its correct shape. The surface finish of the extruded profile must also be very good, as it will often be used in a highly visible part of the car where aesthetic appearance is highly important. Once cured, besides possessing the usual key physical properties, such as good permanent set, good ageing and ozone resistance and a wide operational temperature range, the weatherstrip profile must exhibit good 'snap' or resilience to enable it to cling to adjacent glass or metal to prevent wind, water and travel noise from entering the vehicle [4].

It is not unusual for a material specification, such as for weatherstrip rubber, to be based solely on physical and mechanical performance requirements which could be achieved by any suitable rubber formulation, although price considerations subsequently become important. It is usual to classify weatherstrip material in terms of hardness. Whilst the use of highly filled EPDM alone in automotive rubber profiles may fulfil material specification requirements, the weatherstrip may still have poor 'snap' and a high modulus, both of which reduce ease of installation [5] during vehicle manufacture and parts replacement. Some EPDM profiles are also prone to discolouration once mounted on the car and exposed to direct sunlight [2]; such a visual impact is clearly undesirable. Thus, while the position of EPDM is very strong in the profile market, there is room for improvements to be made. This chapter describes how a long sought-after goal [6], of combining the desirable features of NR (high green strength and good physical properties, especially resilience) with those of EPDM (good weathering and ozone resistance), can be achieved to produce a versatile elastomer blend for use as extruded weatherstrip profile.

It has already been demonstrated in Chapter 14 that EPDM can be modified with the sulphur donors *bis*-alkylphenol disulphide (BAPD) and dithiodicaprolactam (DTDC) as part of a normal masterbatch mixing cycle in an internal mixer. EPDM modified in a 'reactively mixed' procedure can be cross-blended satisfactorily with NR to give a blend with significantly improved physical properties compared with a similar blend containing unmodified EPDM. Such an improvement in physical properties is a result of an increased level of crosslinking in the EPDM phase, an increased level of interaction between EPDM and carbon black and an improved blend phase morphology, all of which are brought about by EPDM modification. As in the work on NR/EPDM blends for tyre sidewalls (Chapter 15), a 60:40 NR:EPDM blend ratio has been adopted to provide ozone resistance without the need for added antiozonant [7].

## 17.2 GENERAL EXPERIMENTAL PROCEDURES

The approach to the preparation of NR/EPDM profile compounds has been essentially the same as described in Chapter 14; firstly to prepare separate NR and EPDM masterbatches, then to cross-blend the two masterbatches using the ratio 60:40 NR:EPDM. The NR and EPDM grades used were SMR L and combinations of Polysar 585 and Polysar 5875, respectively. Both of these EPDM grades have a high level of ENB unsaturation and the latter is a high molecular weight, oil-extended (100 phr) grade.

Small-scale masterbatch and cross-blend mixing was performed in a Brabender PL2000E fitted with a 350S mixing head and Banbury style rotors (300 ml capacity, 0.75 fill factor) or a 00C Banbury (3 litre capacity, 0.75 fill factor) internal mixer. Finalizing of small-scale mixes was carried out on a two-roll mill in the usual manner. Press curing was achieved in an electrically heated press at either 160 °C or 180 °C.

Larger scale mixing has been carried out in a variety of internal mixers; a K2A Francis Shaw Intermix having intermeshing rotors (44 litre capacity, 0.64 fill factor), a F 50 Farrel mixer fitted with Synchronous Technology (ST) rotors (47 litre capacity, 0.72 fill factor) and a GK-70 Werner-Pfleiderer internal mixer (70 litre capacity, 0.75 fill factor). The last mixer was used to prepare compound for larger scale extrusion and microwave curing trials. For the GK-70, modified EPDM masterbatches were typically mixed at a rotor speed of 40 rpm to an indicated temperature of 160 °C before dumping, the temperature of the dumped batch being ~175 °C. This procedure would allow the batch to reach a temperature of at least 150 °C and for this to be maintained for about 1.5 minutes to allow the modification reaction to occur to a satisfactory level. The total mixing time of the masterbatches is about 3.5 to 4.0 minutes. Cross-blending is typically achieved using a two-minute mix cycle at a medium rotor speed. Finalizing with curatives is performed either in the internal mixer or on a two-roll mill, employing an efficient vulcanization (EV) system with a combination of primary and secondary accelerators to avoid bloom.

Extrusion of the unvulcanized compounds was performed on a small scale using laboratory size 30 mm, 40 mm and 60 mm Francis Shaw single screw extruders fitted with automatic barrel, screw (40 and 60 mm extruder only) and die head temperature control. On a larger scale, for microwave curing trials, a 90 mm single screw cold feed vacuum extruder equipped with a heating/cooling system with automatic temperature control for screw, barrel and die head has been used. The extrusion performance of compounds mixed on a smaller scale was assessed using a Garvey die [8]. For the larger scale mixing trials either a demonstration die, specially shaped to produce extrusions having a combination of relatively flat surfaces, corners and thin sections, or a typical commercial weatherstrip die, acquired from a Malaysian manufacturer, was employed. The extrudate gradings/ratings used were based on the ASTM method employing Extrusion Die No. 1, Garvey type [8]. Key physical properties of vulcanizates were also determined; these included hardness, tensile strength, elongation at break, compression set, ageing properties and ozone resistance.

Continuous curing of extrudates was achieved using a microwave oven and hot air tunnel system. The cured extrudate was passed through a cooling trough and taken up on a conveyor fitted with cutting equipment. The microwave oven used has an available power of 10 kW: the energy is generated by four magnetrons with a nominal rating of 3 kW each, but which are used at a maximum of 2.5 kW. The power can be infinitely adjusted up to 10 kW. The tunnel has a total of six cavities with the first and last cavities being empty, i.e. not fitted with a magnetron. The microwave and hot air tunnels were set at temperatures of 200 °C and 180 °C respectively.

## 17.3 INITIAL PREPARATION AND ASSESSMENT OF BLEND COMPOUNDS

A recent material specification [9] from the Ford Motor Company for EPDM weatherstrip gives an indication (Table 17.1) of the properties required for a range of different hardness compounds. This provides a starting point at which to aim in an initial assessment of the suitability of the EPDM modification procedure for use in an extruded profile NR/EPDM blend. Formulations for representative NR and EPDM control compounds were derived from the MRPRA Formulary [10] and a pilot production formulation [11] from a major European EPDM manufacturer, respectively. Minor adjustments were made, however, in the light

**Table 17.1** Specifications for commercial weatherstrip [9]

| *Original property* | *A1 Soft* | *A2 Medium soft* | *A3 Medium hard* | *A4 Hard* |
|---|---|---|---|---|
| Hardness, IRHD | 55–65 | 65–75 | 75–85 | 85–95 |
| Tensile strength, MPa | >6.0 | >6.0 | >6.0 | >7.0 |
| Elongation at break, % | >250 | >225 | >150 | >120 |
| M100, MPa | 1.5–4.5 | 2.0–5.0 | 3.5–7.0 | 5.0–8.5 |
| Tear resistance (Die C), kN/m | >18 | >18 | >18 | >18 |
| Rebound resilience, % | >25 | >25 | >25 | >25 |
| Compression set, 22 h/70 °C | <35 | <35 | <35 | <35 |

of materials availability. The formulations used are detailed in Tables 17.2 and 17.3.

An initial NR/EPDM blend formulation was devised (Table 17.4) based on the two single polymer formulations noted above. As in the previously described work on NR/EPDM blends (Chapter 14), a separate masterbatch and cross-blending procedure was adopted to facilitate modification of only the EPDM phase of the blend. Modification of EPDM in the internal mixer is the most important step of the whole blend preparation procedure. The batch ingredients are added early on in the mix cycle after which the ram remains down for the remainder of the mix cycle. It is essential that the masterbatch itself reaches a temperature of

**Table 17.2** NR control formulation [10]

| *Ingredient* | *phr* |
|---|---|
| SMR 20 | 100 |
| N762 | 65 |
| Struckpar 2280 | 5 |
| Struktol WB 212 | 4 |
| TMQ | 2 |
| Wingstay 100 | 0.5 |
| Sunproof Improved wax | 7 |
| Zinc oxide | 5 |
| Stearic acid | 1 |
| Sulphur | 1.2 |
| MBTS | 2 |
| ZDBC | 0.5 |

**Table 17.3** EPDM control formulation

| *Ingredient* | *phr* |
|---|---|
| Keltan 512 × 50 | 150 |
| N550 | 140 |
| Whiting | 50 |
| Flexon 876 oil | 40 |
| Zinc oxide | 3 |
| Stearic acid | 1 |
| Sulphur | 1.2 |
| MBT | 1.2 |
| TMTD | 0.64 |
| DPTH | 1 |
| ZDBC | 0.64 |

**Table 17.4** NR/EPDM blend formulation

| *Ingredient* | *phr* | |
|---|---|---|
| SMR 20 | 60 | 60 |
| Polysar 585 | 40 | 40 |
| BAPD[a] | – | 0.32 |
| DTDC[a] | – | 0.16 |
| N762 | 80 | 80 |
| Struckpar 2280 | 12 | 12 |
| Struktol A50 | 2 | 2 |
| Zinc oxide | 5 | 5 |
| Stearic acid | 1 | 1 |
| TMQ[b] | 0.5 | 0.5 |
| Sulphur | 0.5 | 0.5 |
| MBTS | 1.5 | 1.5 |
| TMTD | 1 | 1 |
| ZDBC | 0.4 | 0.4 |

[a] Added to EPDM masterbatch only
[b] Added to NR masterbatch only

approximately 160 °C and that it is maintained at this temperature for one to two minutes (dependent upon black type and loading) before dumping. It is important that the batch should not be mixed beyond the point where any significant upturn in mixing torque is observed, as this is an indication of the development of an excessive level of interaction between the EPDM and carbon black. The extent of EPDM modification was quantified by GPC/UV analysis of an extracted portion of the EPDM masterbatch as described in Chapter 14.

Compared with the NR control compound, an increased black and oil loading was used in the blends, although not to the extent used in the EPDM control formulation. Some adjustment of the cure system was made to take account of the NR/EPDM ratio and the slow rate of vulcanization of EPDM in comparison to NR. The level of sulphur was reduced, to increase the efficiency of vulcanization and to improve the ageing resistance of the NR phase in the blend. A low level of antioxidant was added only to the NR masterbatch before cross-blending.

The effect of the BAPD/DTDM modification of EPDM on the phase morphology of the vulcanized blend was studied by TEM (Figures 17.2 and 17.3). Whilst it has been demonstrated in an unfilled blend that modification of EPDM can have a considerable effect upon phase morphology (Chapter 14), it is more difficult to discern clearly the phase morphology of black-filled blends. However, the unmodified blend (Figure 17.2) does appear to contain regions largely devoid of carbon black filler, seen as very light areas, which are not seen in the NR/

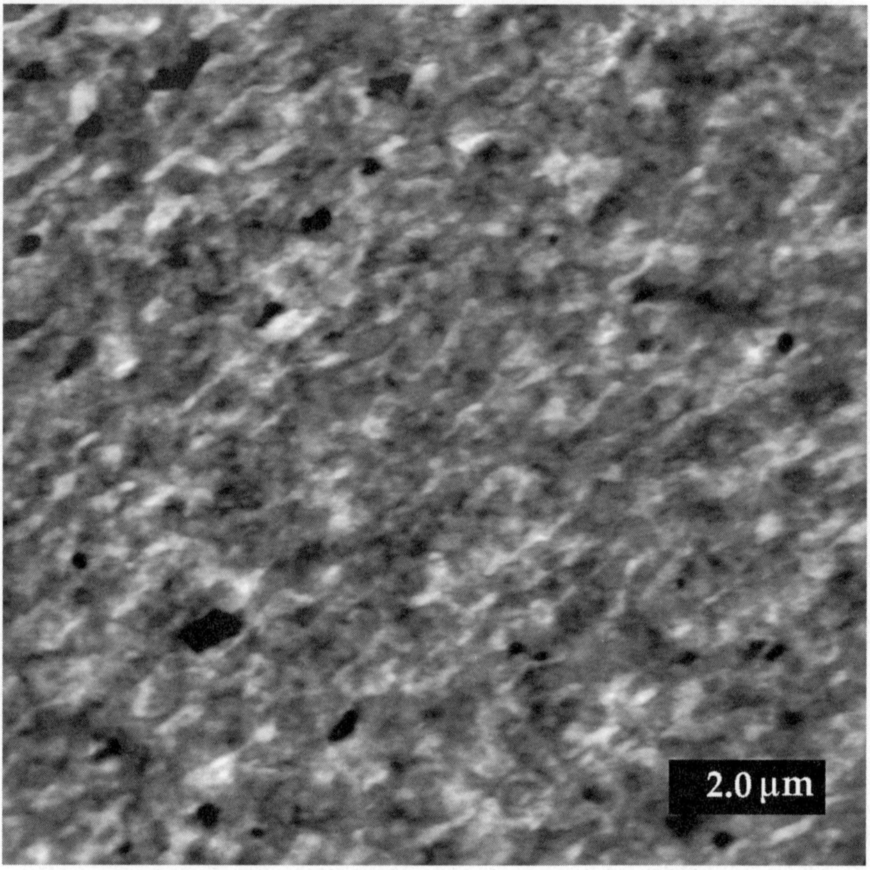


**Figure 17.2** TEM micrograph of a black filled 60:40 NR:unmodified-EPDM blend.

modified-EPDM blend (Figure 17.3). This is not unexpected considering the effectiveness of the EPDM modification in improving the level of interaction between the polymer and carbon black, as indicated by the large increase in bound rubber content [12] of the modified EPDM masterbatch (Table 17.5). The NR/modified-EPDM blend also exhibits a more granular or textured appearance which is not shown by the blend containing unmodified EPDM. Such an improved phase morphology has become a recognizable feature of NR/modified-EPDM blends. An improved blend phase morphology may also be expected to improve the resistance of the blend to ozone attack, although ozone resistance of all of the 60:40 NR/EPDM blends is adequate, i.e. no cracking being observed after three days at 20% strain at an ozone concentration of 50 pphm.

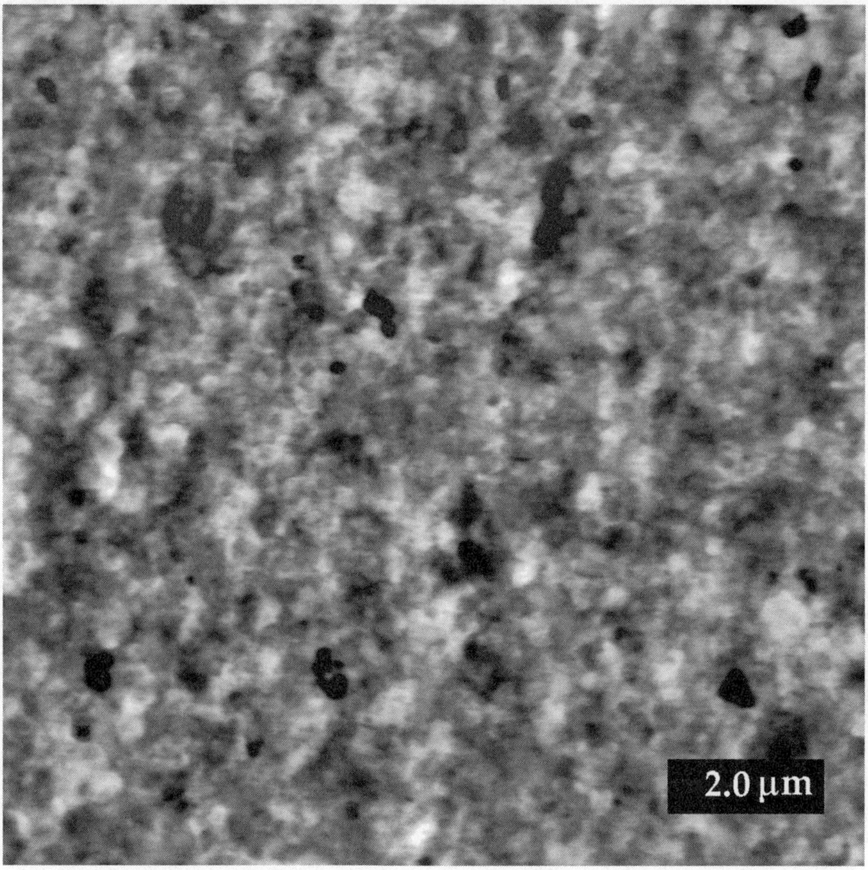


**Figure 17.3** TEM micrograph of a black-filled 60:40 NR:(BAPD/DTDC) modified-EPDM blend.

Compound and vulcanizate properties of the NR and EPDM control compounds and NR/EPDM blends are shown in Tables 17.5 and 17.6. The results indicate that the EPDM modification procedure is effective in producing appreciable improvements to the tensile properties of the blend. Retention of tensile properties on ageing (seven days at 70 °C) is shown to be in most cases better than that of the NR control and is a considerable improvement in performance over the EPDM control compound. Compression set of the blend is not as good as that of the EPDM control compound in performance, although the level of improvement sought is not great and should be attainable.

The EPDM control and NR/EPDM blend vulcanizates all displayed a surface bloom which was absent from the NR vulcanizate. Examination

**Table 17.5** Compound properties of the first series of NR/modified-EPDM blends

| *Material* | *Rheometry at 160° C* | | | *Mooney viscosity*[a] | *Bound modifier in* | *Bound rubber, g/g black* | |
|---|---|---|---|---|---|---|---|
| | $t_{s2}$ *min:sec* | $t_{95}$ *min:sec* | *Torque rise (dNm)* | *ML (1 + 4) 100 °C* | *EPDM m/b, wt%* | *NR phase* | *EPDM phase* |
| NR control | 1:18 | 3:04 | 10.5 | 42 | – | – | – |
| EPDM control[b] | 0:45 | 4:10 | 13.7 | 77 | – | – | – |
| NR/EPDM | 1:41 | 9:21 | 13.1 | n/a | – | 0.24 | 0.16 |
| NR/mod-EPDM | 1:30 | 7:59 | 13.5 | 59 | 0.303 | 0.22 | 0.41 |

[a] Prior to finalizing
[b] Rheometry at 180 °C

of the bloom by thin layer chromatography showed it to contain ZDMC, which is formed from TMTD during vulcanization. The formation of a surface bloom in EPDM compounds when using thiuram accelerators is not an uncommon problem [2], and can be solved by reducing the TMTD or partly replacing it by a low level of another thiuram accelerator.

## 17.4 SMALL-SCALE EXTRUSION TRIALS

Two NR/modified-EPDM blends together with NR and EPDM control compounds similar to those described above were prepared on a larger scale using a 00C Banbury internal mixer (approximate batch volume 3 litres). The blend compounds were mixed in order to demonstrate that the modification of EPDM using sulphur donors could be achieved satisfactorily in a larger sized mixer than a small laboratory mixer. Mixing on a larger scale also enables sufficient material to be prepared for an assessment of the extrusion performance of the blends on a laboratory scale 30 mm single screw extruder.

Formulations for these compounds is given in Tables 17.2 and 17.3 for the control compounds and Table 17.7 for the blend compounds. A minor change was made to the NR control compound formulation given in Table 17.2 in that the 65 phr of N762 carbon black was replaced by the same quantity of N550. The two NR/modified-EPDM blend compounds were prepared with low and medium levels of N660 black in the NR masterbatch and medium and high levels of black in the EPDM masterbatches. Rather than Polysar 585 being used on its own, a mixture of Polysar 585 and Polysar 5875 were used. The incorporation of a higher

**Table 17.6** Physical properties of the first series of NR/EPDM blend vulcanizates[a]

| *Material* | *M100 (MPa)* | *TS (MPa)* | *EB (%)* | *Hardness (IRHD)* | *Tear[b] strength (N/mm)* | *Compression set (%)* | *Tension set[c] (%)* | *Dunlop resilience[d] (%)* |
|---|---|---|---|---|---|---|---|---|
| NR control | | | | | | | | |
| Unaged | 1.87 | 20.9 | 545 | 58 | 35.6 | 33 | 5.9 | ~70 |
| Aged, % change | 25 | −7.7 | 15.8 | 2.6 | – | – | – | – |
| EPDM control | | | | | | | | |
| Unaged | 4.04 | 12.6 | 340 | 68 | 7.5 | 23 | 3.2 | 47.3 |
| Aged, % change | 50.2 | 7.9 | −23.5 | 6.7 | – | – | – | – |
| NR/unmodified-EPDM | | | | | | | | |
| Unaged | 2.85 | 10.7 | 380 | 67 | 6.1 | 25 | 6.6 | ~60 |
| Aged, % change | 18.2 | −1.9 | −13.2 | 1.5 | – | – | – | – |
| NR/modified-EPDM | | | | | | | | |
| Unaged | 2.88 | 13.8 | 425 | 66 | 7.1 | 25 | 7.2 | ~60 |
| Aged, % change | 14.2 | 0 | −8.2 | 3 | – | – | – | – |

Ageing conditions, 7 day/70 °C, values expressed as percentage change
[a] Vulcanizates cured to $t_{95}$ at 160 °C
[b] Trouser tear
[c] 22 h at 100% strain, 30 min recovery
[d] Approximate resilience values estimated from properties of similar formulation compounds
[e] Vulcanizates cured to $t_{95}$ at 180 °C

**Table 17.7** Formulations of NR/modified-EPDM blends for initial extrusion trials

| *Ingredient* | *Mix no.* | |
|---|---|---|
| | *076 (phr)* | *077 (phr)* |
| EPDM masterbatch[a] | | |
| Polysar 585 | 80 | 60 |
| Polysar 5875 | 40 | 80 |
| N660 | 60 | 120 |
| Struckpar 2280 | 3 | 6 |
| Struktol A50 | 5 | – |
| NR masterbatch[b] | | |
| SMR 20 | 100 | 100 |
| N660 | 30 | 50 |
| Struckpar 2280 | 2.5 | – |
| Cross-blend | | |
| Overall black | 52 | 78 |
| Overall oil | 10.1 | 19.9 |

Common to all masterbatches: 3 phr zinc oxide and 1 phr stearic acid. Cure system, phr: sulphur 0.25; MBTS 1.5; TMTD 0.25; TBTD 0.5; DPTH 0.2; ZDBC 0.8.
[a] EPDM masterbatch modified with 0.8 phr BAPD and 0.4 phr DTDC
[b] 3.33 phr Struktol A50 and 0.83 phr TMQ added to NR masterbatch

molecular weight, oil-extended EPDM with a lower molecular weight non oil-extended EPDM was expected to lead to a higher overall EPDM crosslink density, increased interaction with carbon black and better cross-blending between masterbatches (Chapter 15). The proportion of the two EPDM grades was altered according to the levels of black in the masterbatches so that only a small proportion of additional free oil was required. The mixing procedure adopted for EPDM modification was essentially the same as described above.

The rheometer characteristics of the compounds at 180 °C are presented in Table 17.8. It can be seen that the NR control has less scorch safety than the blends or the EPDM control compound. As might be expected, the NR control is also much faster curing than the blends which reach $t_{95}$ in about half the time of the EPDM control. In practice it would be reasonable to cure an EPDM profile at perhaps 20° to 40 °C higher temperature. The torque rise shown by the blends is much lower

**Table 17.8** Rheometer[a] characteristics of compounds for initial extrusion trials

| *Mix* | $t_{s2}$ *min:sec* | $t_{95}$ *min:sec* | *Torque rise (dNm)* |
|---|---|---|---|
| NR control | 0:31 | 0:55 | 10.5 |
| EPDM control | 0:44 | 4:50 | 14.4 |
| NR/EPDM 076 | 0:47 | 2:30 | 7.03 |
| NR/EPDM 077 | 0:41 | 2:10 | 8.69 |

[a] Monsanto MDR 2000E at 180 °C

than that of either of the controls and reflects to some extent the comparatively low state of cure attained by the EPDM phase in the blends.

Extrusion performance of the above four compounds was determined using a Francis Shaw 30 mm extruder fitted with a Garvey type die [8]. The range of angles and thicknesses produced by a Garvey die are typical of those found in complex commercial weatherstrip profiles. The extrusion conditions used (die head and barrel temperature) to prepare the four extrudates were the same for all of the compounds. Photographs of the four extrudates are shown in Figure 17.4.

Table 17.9 shows how the compounds compare in the different categories of extrudate assessment. It can be seen both from the assessments and the photographs that the extrusion performance of one of the blends is not as good as that of the control compounds. There appeared to be nibs of 'cured' material throughout the profile from mix No. 077 (high black loading in EPDM, medium black loading in NR). The appearance of such nibs is thought to be associated with highly gelled regions of the EPDM arising from an excessive level of coupling to carbon black developed during the modification procedure. This problem would be exacerbated by high black loadings in the EPDM phase and a high proportion of high molecular weight Polysar 5875. This is evidenced by the good extrusion performance of compound No. 076, which has a medium black loading in EPDM (lower proportion of Polysar 5875) and a medium black loading in NR.

It was apparent from the nature of the extrudates that both of the NR/modified-EPDM blends possessed a very high green strength. This is thought to result from the high level of coupling developed between the EPDM and carbon black during the reactivity mixed EPDM masterbatch stage rather than from a lack of processing safety, since the NR control compound exhibited a shorter scorch period and a faster rate of cure than either of the blend compounds. The NR control compound also showed good green strength, but to a lesser degree than the blends. In contrast,

**Figure 17.4** Photographs of Garvey die extrudate [8].

**Table 17.9** Garvey extrusion ratings [8] of compounds for initial extrusion trials

| *Mix* | *Swelling–porosity* | *Extrusion gradings* | | |
|---|---|---|---|---|
| | | *Sharpness and continuity of 30° edge* | *Smoothness of surface* | *Sharpness and continuity of corners* |
| NR control | 3–4 | 4 | 4 | 4 |
| EPDM control | 4–4 | 4 | 4 | 4 |
| NR/EPDM 076 | 3–4 | 4 | 4 | 4 |
| NR/EPDM 077 | 3–3 | 3 | 2 | 3 |

Rating 1 (poor) to 4 (excellent)

the EPDM control compound possessed relatively poor green strength. As indicated above, a high green strength could be seen as commercially advantageous as it is important for an extruded profile to be able to support its own weight and maintain its shape prior to high temperature curing.

## 17.5 FORMULATION DEVELOPMENT AND PROCESSING

The feasibility of mixing these modified EPDM masterbatches on a larger scale has been demonstrated in a K2A Francis Shaw Intermix having intermeshing rotors (44 litre capacity, 0.64 fill factor) and a Farrel F50 Banbury mixer fitted with Synchronous Technology (ST) rotors (47 litre capacity, 0.72 fill factor). Modification of an EPDM (70:60 Polysar 585:Polysar 5875) masterbatch containing 70 phr N550 carbon black can be achieved in a reasonable mixing time provided that efficient mixing is maintained throughout the cycle. Computer control of the mixing cycle, which is commonplace on larger mixers having variable speed drives, will further facilitate the ease of EPDM modification as control of the batch temperature can be maintained through manipulation of the rotor speed. Processing of the masterbatch from the mixer can be achieved in the usual manner on a mill or by using a dump extruder. However, it should be noted that continuation of the modification process is possible if the batch remains at an elevated temperature for an extended period and excessive levels of interaction between the EPDM and carbon black can develop.

For a range of physical properties, including hardness, tensile strength, compression set and resilience, the most appropriate modifier combination for modification of the EPDM masterbatch has been shown to be 0.8 phr BAPD with 0.4 phr DTDC. This was determined using designed experiment methods [13, 14] with compounds prepared on a laboratory scale using a 00C Banbury mixer (3 litre capacity). Whilst modification of EPDM using sulphur donors has been demonstrated to give improvements in a broad range of physical properties, compound extrusion performance, as determined using a Garvey die, was considered inferior to those of the NR and EPDM controls, particularly in respect of the sharpness and continuity of the 30° edge. Die design has already been shown (Chapter 15) to be crucial in producing extrudate of acceptable quality and, whilst the deficiency of the NR/EPDM compound with regard to extrudate swell is a problem, it would be expected to be solved by the development of a suitable die.

Material for use in extrusion and microwave curing trials has been prepared on a large scale using a GK-70 Werner-Pfleiderer internal mixer (70 litre capacity, 0.75 fill factor). Table 17.10 shows the preferred basic formulation used to prepare the NR/EPDM masterbatches. Two com-

**Table 17.10** NR and EPDM masterbatch formulations for large-scale mixing and extrusion trials of soft and medium-soft blend compounds

| *EPDM masterbatch* | *Soft (phr)* | *Medium-soft (phr)* |
|---|---|---|
| Polysar 585 | 80 | 50 |
| Polysar 5875 | 40 | 100 |
| N550 black | 60 | 100 |
| Sunpar 2280 oil | 3 | 15 |
| Strucktol WB212 | 5 | 5 |
| BAPD | 0.8 | 0.8 |
| DTDC | 0.4 | 0.4 |
| Zinc oxide | 3 | 3 |
| Stearic acid | 1 | 1 |
| NR masterbatch | | |
| SMR L | 100 | 100 |
| N550 black | 30 | 50 |
| Sunpar 2280 oil | 1.5 | 5 |
| Zinc oxide | 3 | 3 |
| Stearic acid | 1 | 1 |
| TMQ | 0.85 | 0.85 |
| Cross-blend | | |
| Overall N 550 | 42 | 70 |
| Overall oil | 10.1 | 29 |

pounds of different hardness were prepared, a soft and medium soft compound. The soft compound is aimed at thin, small sponge profiles, e.g. for car door liners. The medium soft range is aimed at thicker, solid profiles, e.g. for van rear window seals. The curing system used for the first set of formulations was 0.6 phr sulphur, 1.0 phr TBBS, 1.0 phr TMTD, 0.25 phr DPTH, 0.5 phr TBUT and 0.8 phr ZDBC. These mixes are designated as BM108/1 and BM108/2.

Table 17.11 shows the extrusion performance for both compounds (mixes 1 and 2) using the demonstration die and 40 mm extruder. The results show that both compounds extrude to give satisfactory surface smoothness and sharpness and continuity of corners, but show a high degree of swell. Porosity was also observed and the effect was worst when the extrudates were vulcanized using the microwave/hot air oven process. Porosity was probably a result of the high humidity in the mixing and extrusion areas. To solve this problem, 5 phr of Rhenosorb C/GW (an anti-moisture agent) was incorporated into both compounds prior to extrusion. The results in Table 17.11 show that the inclusion of an

**Table 17.11** Extrusion ratings of large-scale mixing trial compounds

| *Mix no.* | *Porosity* | *Swelling* | *Surface smoothness* | *Sharpness and continuity of corners* |
|---|---|---|---|---|
| Without anti-moisture agent | | | | |
| BM108/1 | 2 | 2 | 3–4 | 3–4 |
| BM108/2 | 2–3 | 3 | 3–4 | 3–4 |
| With anti-moisture agent | | | | |
| BM108/3 | 3 | 4 | 4 | 4 |
| BM108/4 | 3–4 | 4 | 4 | 4 |
| EPDM control | 4 | 4 | 4 | 4 |
| (Commercial compound) | | | | |

Ratings 1 (poor) to 4 (excellent), extruded using a demonstration profile die

anti-moisture agent leads to extrudates which exhibit significantly less swelling and porosity. For comparative purposes, a control EPDM compound obtained commercially (formulation unknown) showed excellent extrusion performance. The dimensional ability of the latter compounds was also good.

The key physical properties of compounds 3 and 4 prepared with the inclusion of Rhenosorb C/GW are shown in Table 17.12. The values are compared with the soft and medium-soft compounds in the Ford specification [9]. The results indicate that the blend compounds produced have hardness values which correspond to just below the lower limits of the soft and medium-soft specifications. The softer blend exhibited better tensile properties than the harder grade. It is also noteworthy that all

**Table 17.12** Physical properties of large-scale mixing trial compounds

| *Property* | *Commercial specification [7]* | *NR/modified-EPDM* | |
|---|---|---|---|
| | | *BM108/3* | *BM108/4* |
| Hardness, IRHD | 55–65 | 53 | 63.5 |
| TS, MPa | >6.0 | 17.7 | 13.9 |
| EB, % | >250 | 585 | 385 |
| Compression set, % 1 day/70 °C | <35 | 28 | 25 |
| Resilience, % | >25 | 72 | 62 |
| Ozone resistance (50 pphm/72 hr/20%/40 °C) | No cracks | No cracks | No cracks |

properties of the blends are significantly better than the specified limits. Ozone resistance after three days at 20% strain in 50 pphm ozone is adequate; there is no cracking.

Tensile properties of the NR/modified-EPDM blend vulcanizates after ageing in hot air are presented in Table 17.13. The tensile strength of the blends holds very well, particularly for the harder compound, although there was slight stiffening as indicated by the increases in hardness. The percentage change in elongation at break after ageing was also slightly higher for the harder compound.

Although the specification states that the maximum allowable compression set is 35%, contact with a manufacturer indicated that compression set values lower than 25% would be preferred. A blend formulation shown in Table 17.14, chosen for improved compression set, was mixed in a GK-70 internal mixer employing similar mixing cycles and procedures to those described earlier. The level of modifying agent added to the EPDM masterbatch was reduced by 25% to 0.6 phr BAPD/ 0.3 phr DTDC to reduce the risk of the level of interaction between the EPDM and carbon black from becoming too high and adversely affecting subsequent processing.

It has been shown that increasing the level of sulphur to 0.85 phr and replacing the TMTD with 1.5 phr MBTS improves compression set marginally. Little difference was observed when the DPTH was omitted and the ZDBC was replaced by an increased level of MBTS. Based on this information, two curative systems were then selected in finalizing the cross-blend masterbatches (Table 17.15). Rhenosorb C/GW was also incorporated in the finalized blend at 5 phr.

**Table 17.13** Resistance of large-scale mixing trial compounds to ageing in air for 7 days/70 °C

| *Property* | *NR/modified-EPDM* | |
|---|---|---|
| | *BM108/3* | *BM108/4* |
| Hardness, IRHD | | |
| Unaged | 53 | 63.5 |
| Aged | 58 | 69 |
| TS, MPa | | |
| Unaged | 17.7 | 13.9 |
| Aged | 14.5 | 13.5 |
| EB, % | | |
| Unaged | 585 | 385 |
| Aged | 420 | 270 |

**Table 17.14** Formulations of NR and EPDM masterbatches for improved compression set profiles

| *Ingredient* | *NR masterbatch (phr)* | *EPDM masterbatch (phr)* |
|---|---|---|
| SMR 20 | 100 | – |
| Polysar 585 | – | 70 |
| Polysar 5875 | – | 60 |
| N550 Black | 50 | 70 |
| BAPD | – | 0.6 |
| DTDC | – | 0.3 |
| Sunpar 2280 oil | 5 | 10 |
| Strucktol WB212 | 1 | 5 |
| Zinc oxide | 5 | 5 |
| Stearic acid | 2 | 1 |
| TMQ | 0.85 | – |

**Table 17.15** Effect of cure system on compression set of NR/modified-EPDM blends

| *Curative* | *Mix no.* | |
|---|---|---|
| | *BM108/5* | *BM108/6* |
| Sulphur | 0.85 | 0.85 |
| TBBS | 2.3 | 2.3 |
| TMTD | 1.2 | – |
| MBTS | – | 2.0 |
| DPTH | 0.29 | 0.29 |
| TBTD | 0.58 | 0.58 |
| ZDBC | 1.15 | 1.15 |
| Compression set, % | | |
| 3 days/23 °C | 23 | 22 |
| 1 day/70 °C | 20 | 21 |

The results listed in Table 17.15 show that the incorporation of either TMTD or MBTS at the levels specified in the blends reduces compression set to around 22%, which is marginally higher than for the EPDM control (acquired commercially) which has a compression set of 21%. Further improvement should not be necessary. Of the other key physical properties, it was found that the NR/modified EPDM blends exhibited significantly better tensile properties compared to the EPDM control

**Table 17.16** Physical properties of NR/modified-EPDM blend vulcanizates

| | *EPDM control* | *Mix no.* | |
|---|---|---|---|
| | | *BM108/5* | *BM108/6* |
| Hardness, IRHD | 65.5 | 65.5 | 63.5 |
| Tensile strength, MPa | 9.5 | 13.1 | 15.1 |
| Elongation at break, % | 350 | 340 | 400 |
| Crescent tear, N/mm | n/a | 21.0 | 27.7 |

**Table 17.17** Extrusion ratings of NR/modified-EPDM blend compounds

| *Mix no.* | *Swelling* | *Porosity* | *Surface smoothness* | *Sharpness and continuity of corners* |
|---|---|---|---|---|
| EPDM control (Commercial) | 4 | 4 | 4 | 4 |
| BM108/5 | 3–4 | 4 | 4 | 4 |
| BM108/6 | 3–4 | 4 | 4 | 4 |

Rating 1 (poor) to 4 (excellent), extruded using a demonstration profile die

(Table 17.16). The MBTS curing system appears to give better tensile strength and elongation at break than the TMTD system.

The extrusion performance of compounds BM108/5 and BM108/6 was assessed on a 90 mm cold feed vacuum extruder using a commercial weatherseal die as described above. The extrudates were then continuously vulcanized through a microwave/hot air oven. The results in Table 17.17 show that the surface finish and dimensional stability of the extrudates were as good as the EPDM control compound. Extrudate swell was slightly higher than that of the EPDM control compound; this would be expected for most compounds containing a significant amount of NR. The high extrudate swell could be reduced by 'drawing down' the extrudate, or in the design of the die.

## 17.6 CONCLUSION

It has been shown that an EPDM masterbatch formulated for use in extruded profiles can be successfully modified with commercially available sulphur donors as part of a normal EPDM mixing cycle in a range of different types and sizes of internal mixer. Cross-blending with a NR masterbatch to give a 60:40 NR:EPDM blend gives a compound which when vulcanized has good physical properties before and after ageing, coupled with satisfactory ozone resistance. Very good extrusion performance is shown by the NR/modified-EPDM blend compound when extruded at high rates through a commercial profile die. Modified blend compound has also demonstrated its suitability for continuous microwave–hot air oven curing at high temperature. A NR/modified-EPDM blend has been shown to have advantages over both an all-NR profile compound and a current commercial EPDM profile compound. A NR/sulphur donor modified-EPDM blend weatherstrip profile formulation, unlike a NR formulation, does not require the use of antiozonants and thus avoids undesirable blooming during the lifetime of the profile. Unlike many EPDM profile formulations, a NR/modified EPDM blend also possesses good green strength in the uncured state and good 'snap' or resilience when cured.

## 17.7 REFERENCES


1 Furminger, S. (1973) Automobile weatherseals. *Rubber in Transport Engineering*, NRPRA, 9/1–9/5.

2 Jourdain, E.P. (1996) Compounding parameters to control the iridescent bloom in EPDM body sealing applications. Paper No. 42, ACS Rubber Division Meeting, Montreal, Canada, May 5–8.

3 Schauder, J.R. and Stella, G. (1992) The unique design latitude of EPDM rubbers. *Kautschuk Gummi Kunststoffe*, **45**(12), 1055–1060.

4 Snappy Vistalon® EPDM compounds TI-51 (1975), *Elastomers Technical Information*, Exxon Chemicals.

5 Jyawook, S. (1987) Automotive body sealing: Design and technology. Paper c, ACS Rubber Division Meeting – Educational Symposium No. 19 (Nontire automative applications), Cleveland, Ohio, USA, October 6–9.

6 Ohwaki, M. (1987) Current requirements and future trends for automotive application of elastomers. Paper g, ACS Rubber Division Meeting – Educational Symposium No. 19 (Nontire automative applications), Cleveland, Ohio, USA, October 6–9.

7 von Hellens, C.W. (1989) Some considerations for the application of EPDM in tire components. Paper No. 40, ACS Rubber Division Meeting, Detroit, Michigan, USA, October 17–20.

8 American Society for Testing Materials (1990) *Standard test methods for rubber property – Extrudability of unvulcanized compounds*, D 2230–90.

9 Ford Motor Company (1995) Engineering Material Specification No. WDS-M2D405-A1. *Ford Automative Safety and Engineering Standards.*

10 MRPRA (1984) Detergent-resistant washing machine seals. *The Natural Rubber Formulary and Property Index*, Malaysian Rubber Producers' Research Association.

11 Personal communication (1984), DSM Elastomers Europe, The Netherlands.

12 Dannenberg, E.M. (1986) Bound rubber and carbon black reinforcement. *Rubber Chemistry and Technology*, **59**, 512–524.

13 Freakley, P.K. (1985) *Rubber Processing and Product Organisation*. Plenum Press, New York.
14 Davies, V.L. (1954) *Design and analysis of experiments*. Oliver & Boyd, London.

# Glossary

ABBREVIATIONS USED IN TABLES AND TEXT

| | |
|---|---|
| 6PPD | N-(1,3-dimethylbutyl)-N′-phenyl-*p*-phenylenediamine |
| BAPD | *bis*-alkylphenol disulphide |
| BGA | Bundesgesundheitsamt (German Federal Health Authority) |
| BR | polybutadiene rubber |
| CBS | N-cyclohexylbenzothiazole-2-sulphenamide |
| $CHCl_3$ | chloroform |
| CR | polychloroprene rubber |
| CTAB | a measure of the surface area of carbon black, and hence particle size |
| DBEEA | di-(butoxy-ethoxy-ethyl)adipate |
| DBPA | dibutyl phthalate absorption, a measure of carbon black structure |
| DHBP | 2,5-dimethyl-2,5-di-*t*-butylperoxy-hexane |
| Dicup | dicumyl peroxide |
| Diolpate 7107 | a commercial blend of polymeric esters |
| DMDPTD | dimethyl diphenylthiuram disulphide |
| DOP | dioctyl phthalate |
| DOTG | di-*o*-tolylguanidine |
| DPG | diphenyl guanidine |
| DPTH | dipentamethylenethiuram hexasulphide |
| DPTT | dipentamethylenethiuram tetrasulphide |
| DTDC | dithio-di-caprolactam |
| DTDM | dithio-di-morpholine |
| DV | dynamic vulcanization |
| ENR | epoxidized natural rubber |
| ENR-25 | epoxidized natural rubber, 25 mole% modification |
| ENR-50 | epoxidized natural rubber, 50 mole% modification |
| EPDM | ethylene propylene diene rubber |
| EPM | ethylene propylene rubber |
| EV | efficient vulcanization, low sulphur:accelerator ratio |
| FDA | US Food and Drugs Administration |
| FT | Fourier-transform |
| G* | complex shear modulus |

| | |
|---|---|
| $G_s$ | static shear modulus |
| GC–MS | gas chromatography–mass spectroscopy |
| GPC/UV | gel permeation chromatography with UV detection |
| HVA-2 | N,N′-m-phenylenebismaleimide |
| Hv-BR | high vinyl-BR |
| IPPD | N-isopropyl-N′-phenyl-*p*-phenylenediamine |
| LDPE | low density polyethylene |
| LM | light microscopy |
| $K_d$ | partition coefficient |
| $M_H$ | maximum torque rise on Monsanto cure rheometer |
| $M_L$ | minimum torque rise on Monsanto cure rheometer |
| MR100 | relaxed modulus at 100% strain |
| MBS | N-oxydiethylenebenzothiazole-2-sulphenamide |
| MBT | 2-mercaptobenzothiazole |
| MBTS | 2,2′-benzothiazole disulphide |
| MG30 | poly(methyl methacrylate) grafted NR (30% PMMA by weight) |
| NBR | acrylonitrile butadiene rubber |
| NMR | nuclear magnetic resonance |
| NR | natural rubber |
| ODIP | N,N′-dioctadecyl N,N′-diisopropylthiuram disulphide |
| PMMA | poly(methyl methacrylate) |
| phr | parts per hundred rubber |
| pphm | parts per hundred million |
| ppm | parts per million |
| Robac $\gamma$ | activated, mixed dithiocarbamates |
| Robac ZIX | activated xanthate accelerator |
| S | sulphur |
| SBR | styrene-butadiene rubber (random copolymer) |
| SEM | scanning electron microscopy |
| STEM | scanning transmission microscopy, here SEM using a modified stage allowing transmission microscopy |
| $ts_1$ | time to 1 dNm rise on a cure rheometer torque |
| $t_{90}$ | time to 90% of torque rise on a cure rheometer |
| $t_{95}$ | time to 95% of torque rise on a cure rheometer |
| $t_{max}$ | time to maximum torque rise on a cure rheometer |
| $T_g$ | glass transition temperature |
| TAC | triallylcyanurate |
| TBBS | N-*t*-butylbenzothiazole-2-sulphenamide |
| TBEP | tributoxyethyl phosphate $(C_4H_9OC_2H_4O)_3PO$ |
| TBSI | N-*t*-butyl-2-benzothiazyl sulphenamide |
| TBTD | tetrabutylthiuram disulphide |
| TEM | transmission electron microscopy |
| TETD | tetraethylthiuram disulphide |

| | |
|---|---|
| TGA | thermogravimetric analysis |
| TMQ | polymerized 2,2,4-trimethyl-1,2-dihydroquinoline |
| TMS | tetramethylsilane |
| TMTD | tetramethylthiuram disulphide |
| TMTM | tetramethylthiuram monosulphide |
| $V_r$ | volume fraction of rubber in a swollen gel |
| Vulkacit 1000 | *o*-tolylbiguanide |
| ZBED | zinc dibenzyldithiocarbamate |
| ZDBC | zinc dibutyldithiocarbamate |
| ZDEC | zinc diethyldithiocarbamate |
| ZDMC | zinc dimethyldithiocarbamate |
| ZDNC | zinc diisononyldithiocarbamate |
| ZEH | zinc-2-ethylhexanoate |
| ZMBT | zinc-2-mercaptobenzothiazole |

# Index